IntechOpen

Supporting open minds since 2005

Panorama of Contemporary Quantum Mechanics - Concepts and Applications
http://dx.doi.org/10.5772/intechopen.75282
Edited by Tuong T. Truong

Contributors
Viktor Gerasimenko, Nicolae Enaki, Masaru Tateno, Jioung Kang, Takuya Sumi, Jose Socorro, Marco Reyes, Carlos Villaseñor Mora, Edgar Condori Pozo, Trong Tuong Truong

First published in London, United Kingdom, 2019 by IntechOpen
IntechOpen is the global imprint of INTECHOPEN LIMITED, registered in England and Wales, registration number: 11086078, 7th floor, 10 Lower Thames Street, London, EC3R 6AF, United Kingdom
Printed in Croatia

British Library Cataloguing-in-Publication Data
A catalogue record for this book is available from the British Library

Additional hard and PDF copies can be obtained from orders@intechopen.com

Panorama of Contemporary Quantum Mechanics - Concepts and Applications
Edited by Tuong T. Truong
p. cm.
Print ISBN 978-1-83962-665-4
Online ISBN 978-1-83962-666-1
eBook (PDF) ISBN 978-1-83962-667-8

For EU product safety concerns:
IN TECH d.o.o., Prolaz Marije Krucifikse Kozulić 3, 51000 Rijeka, Croatia, info@intechopen.com or visit our website at intechopen.com.

Meet the editor

Dr. Tuong T. Truong is a graduate of the "Ecole Centrale des Arts et Manufactures", Paris, France and has a Ph.D in theoretical physics from Columbia University, New York, USA. He has served as an assistant professor of physics at the Free University of Berlin (Germany) and a professor of physics at the University François Rabelais in Tours (France) before moving to the University of Cergy-Pontoise (France), where he is presently an emeritus professor. Throughout his career, he has worked successively in quantum mechanics, integrable quantum field theory in low dimensions, exactly soluble models in two-dimensional statistical mechanics, quantum superintegrable models, and integral geometry for imaging science. He has published around 100 research papers in peer-reviewed journals.

Contents

Preface

Tremendous technological progress has been witnessed in the last few decades and this progress is mainly due to advances made in quantum theory (or quantum mechanics as it was known). Quantum theory was created at the beginning of the twentieth century to decipher the growing number of atomic phenomena. However, quantum theory has introduced many unconventional and non-intuitive concepts, which cannot directly be exploited for technological applications. Over the years, thanks to the contributions of many scientists, the understanding of quantum theory has vastly improved and this has led to many of the present-day discoveries. The aim of this book is to describe some of the development aspects of quantum theory, which may incite or generate further useful applications. After an introductory chapter, we have focused on particular topics in quantum theory, which are discussed in the following chapters of this book.

Chapter 2, by Nicolae A. Enaki, concerns the cooperative spontaneous lasing mechanism and its possible quantum retardation effects. The author considers the effects of cooperative scattering and two-photon resonances on the decay of three-level systems involving non-linear dipole type interactions. These effects occurs in hydrogen- (or helium-) like atoms with cascade transitions, in which scattering is in concurrence with resonance via dipole-forbidden transitions. Also discussed are interferences between single and two quantum collective transitions of inverted radiative emissions, two particle collective decay rates, time dependence of kinetic processes, correlations between radiative emission sources as well as their behaviors at short and long periods under retardation effects.

Chapter 3, by V. I. Gerasimenko, discusses new approaches to the evolution of states of large quantum particle systems by means of marginal correlation operators. It is shown that they are governed by the nonlinear quantum BBGKY hierarchy of equations in the Von Neumann dynamical framework of correlation functions. The non-perturbative solution of the Cauchy problem to this hierarchy of nonlinear evolution equations describes the processes of the creation and the propagation of correlations in large quantum particle systems. Described in detail is the collective behavior of quantum many particle systems by means of a marginal one-particle correlation operator, which is a solution of the generalized quantum kinetic equation with initial conditions, in particular as condensed states.

Chapter 4, by Jiyoung Kang, Takuya Sumi and Masaru Tateno, aims to explain the functional mechanisms of biological macromolecular systems. These mechanisms are due to the electron transfer responsible for the redox regulations and the catalytic reactions for hydrogen metabolism and hydrolysis. Electron transfer induces dramatic rearrangements of electronic structures as well as internal three-dimensional structures, which are crucial for biological functions. Two distinct types of rearrangement triggers are found by two distinct approaches:

- Full *ab initio* quantum mechanics calculations,

- Hybrid *ab initio* and molecular dynamics calculations.

First, redox regulations and catalytic reactions for hydrogen metabolism and hydrolysis by electronic transfer, which are catalyzed by transition metal (4Fe-3S) clusters, are obtained using the first approach. Second, dynamic rearrangements of the electronic structures occurring in the catalytic reaction of RNA-protein complexes have instead emerged from the second approach on hyper parallel supercomputer simulations. Such features are characteristic of the electronic structures in biological macromolecular systems.

Chapter 5, by J Socorro, Marco A. Reyes, Carlos Villasen or Mora and Condori Pozo, presents new aspects of supersymmetric quantum mechanics. This is a theoretic extension of the conventional one-dimensional Schrödinger equation of quantum mechanics *via* the so-called factorization method. Supersymmetry incorporates the fermion-boson symmetry into the theory through the so-called supercharge operators. Although many properties are known, the authors have brought up some features that are susceptible to new insights.

I would like to thank the authors for devoting time and effort to providing high quality texts, which are beneficial to a wide audience in perceiving the realm of quantum theory and its appearance in technology. My appreciations go to Ms Kristina Kardum for her patience and help in guiding this book project through the many steps.

Tuong T. Truong
Laboratoire de Physique Théorique et Modélisation,
University of Cergy-Pontoise, Cergy-Pontoise,
France

Section 1

Introduction

Chapter 1

Introductory Chapter: Panorama of Contemporary Quantum Mechanics - Concepts and Applications

Tuong T. Truong

1. Introduction

Quantum mechanics has been around for more than a century. Since its birth at the beginning of the twentieth century, it has undergone a tremendous growth. But, it is only now that quantum mechanics has emerged in our daily life. This is just a normal evolution for any branch of physics. Take for example, the electromagnetic theory. It came into existence with the stunning work of James Clerk Maxwell in 1865 [1], which predicted the existence of radio waves, and this has led to the tremendous development of electronics throughout the twentieth century in the fields of communication, detection, and transmission of information and data.

Quantum mechanics is the physics of subatomic phenomena, which has remained a mysterious domain for a long time. Its laws have bewildered many because they are quite counter-intuitive. Its development has started with the very concept of "quantization," which entails the absorption as well as the emission of energy in discrete amounts and not continuous as it is usually perceived in classical physics. This milestone principle, established by Max Planck toward the end of the nineteenth century, has started a golden age of discoveries during almost three decades.

2. The present role of quantum mechanics in technologies

Since then, quantum principles have been at the foundations of our day-to-day technologies, such as the transistor, computer chip, LASER, GPS, NMR imaging system, LED lamps, solar cells, etc. to name a few. The working of transistors is based simultaneously on the quantum description of matter, namely, the wave-particle duality and the Heisenberg uncertainty principle, which is inherent to quantum evolution equations. In recent years, with the appearance of Big data, massive exchange of information, and the ensuing cryptography challenges, it becomes necessary to turn to quantum engineering to find a way out to manage these problems. A full array of "quantum" technologies has been initiated since the seminal paper of R. P. Feynman "Simulating Physics with Computers" [2], in which the notion of quantum computer was introduced as it is built on new quantum engineering (quantum electronics/spintronics).

There is no doubt that quantum principles are here to stay and will continue to be at the origin of new technological innovations in coming years. This book is intended to give a first glimpse on a few topics of this fascinating development perspective on the future of the real world and to stimulate research in order to meet the futures challenges. Most urgent is the investigation into the quantum behavior of large systems such as populations of photons or atoms in the regime of Bose condensates in which unexpected properties may arise. In particular, some macroscopic effects may be explained from microscopic levels, thanks to quantum mechanics which governs the evolution rules at the atomic level.

One of the most salient features of modern physics is the inherent existence of hidden symmetry in nature. The discovery of such symmetries is often very fruitful in the sense that it leads to further discoveries and predictions. Strangely enough, even symmetry breaking can also be a source of new phenomena occurrence. This is why since decades, one has sought to make quantum mechanics supersymmetric and the pursuit of supersymmetry in elementary particles is still ongoing these days in large particle accelerators.

Finally, it should be mentioned that quantum mechanics has ushered mankind into the area of fictional reality with the search for the realization and exploitation of the quantum concept of entanglement. In 1935, analyzing the possible outcome of a Gedankenexperiment following the rules of quantum mechanics by Einstein, Podolski, and Rosen has arrived at a paradoxical conclusion, known for a long time as the EPR Paradox [3]. This is because one can predict the value of a dynamical quantity of a system, which has classically nothing to do with a companion system on which measurements are performed. Quantum mechanics coins these systems as "entangled." Nowadays, quantum entanglement has been experimentally demonstrated and considered to be the main ingredient in the working of a quantum computer.

As computing is an exponentially growing activity in science and technology as well as in economics, finance, and management, "classical" computers have reached their limits as far as performance and costs are concerned. Quantum computers which are based on totally new quantum concepts (superposition and entanglement) with their revolutionary capacity for data storage and speed of calculation seem to be the ideal solution to the previous problem. Basic logic gates and circuits are cast and discussed in the language of the so-called quantum bits or q-bits (instead of classical bits) which are their building blocks. They open the way to mastering cyber security in data transfer, Big data mining, and the like.

Thus, our future appeared to be structured by quantum principles via quantum technology and engineering, and this book offers a view of what might be the coming reality we will have to deal with.

Author details

Tuong T. Truong
University of Cergy-Pontoise, Cergy-Pontoise, France

*Address all correspondence to: tuong.truong@u-cergy.fr

References

[1] Maxwell JC. A dynamical theory of the electromagnetic field. Philosophical Transactions of the Royal Society of London. 1865;**155**:459-512

[2] Feynman RP. Stimulating physics with computers. International Journal of Theoretical Physics. 1982;**21**(6/7):467-488

[3] Einstein A, Podolsky B, Rosen N. Can quantum-mechanical description of physical reality be considered complete? Physical Review. 1935;**47**(10):777-780

Section 2

Many Particle Quantum Features

Chapter 2

Cooperative Spontaneous Lasing and Possible Quantum Retardation Effects

Nicolae A. Enaki

Abstract

The collective decay effects between the dipole-active three-level subsystems in the nonlinear interaction with dipole-forbidden transitions, like $2S-1S$ of hydrogen-like radiators, are proposed, taking into consideration the cooperative exchanges between two species of atoms through the vacuum field in the scattering and the two-photon resonance processes. One of them corresponds to the situation when the total energy of the emitted two photons by the three-level radiator in the cascade configuration enters into the two-photon resonance with another type of dipole-forbidden transitions of hydrogen-like (or helium-like) atoms. The similar situation appears in the cooperative scattering between two species of quantum emitters when the difference of the excited energies of the two dipole-active transitions of the three-level radiators is in the resonance with the dipole-forbidden transitions of the Hydrogen-like radiators. These effects are accompanied by the interference between single- and two-quantum collective transitions of the inverted radiators from the ensemble. The two-particle collective decay rate is defined in the description of the atomic correlation functions taking into consideration the phase retardation between them. The kinetic equations which describe the cooperative processes as the function of time and correlation are obtained. The behavior of the system of radiators at short and long time intervals in comparison with the retardation time between them is studied.

Keywords: 42.50.Fx Cooperative phenomena in quantum optical systems, 32.80.Qk Coherent control of atomic interactions with photons, 03.65.Ud Entanglement and quantum nonlocality, 03.65.Yz Decoherence, open systems, quantum statistical methods
2000 AMS Subject Classification: Primary 82C10, 81Q15; Secondary 20G42, 81R15

1. Introduction

The single-photon cooperative emission of the inverted system of radiators proposed by Dicke [1] opens the new possibilities of this phenomenon in the description of decay processes in the multilevel system [2] and multi-photon interaction of radiators with EMF (see, e.g., [3, 4]). The experimental possibilities [3, 4] of nonlinear cooperative interaction of radiators with vacuum field remain in the center of attention of many theoretical models proposed in the last time [5, 6].

For example, using the classical and quantum approaches in Refs. [7–10], it is given the quantitative description of two-color super-fluorescence, observed in [2]. In the recent experiment [11], the cooperative emission of excited atomic oxygen relatively the transition $3p^3P \rightarrow 3s^3S$ at wavelength $845nm$ as a result of two-photon photolysis of atmospheric O_2 followed by two-photon excitation of atomic oxygen by a laser pulse at $226nm$ is demonstrated.

Combining single- and two-photon processes, this chapter aims to investigate the cooperative emission of the inverted system of radiators taking into account the resonance between one- and two-photon cooperative transitions of two three-level atomic subsystems represented in **Figure 1**. In this approach, the two dipole-active species of radiators studied in Refs. [12, 13] are replaced with one three-level atomic subsystem Ξ (or V) inverted relative to the single-photon emission in the resonance with $2S$ - $1S$ dipole-forbidden transitions of hydrogen (or He)-like sub-ensemble. This new cooperative effect between two species of radiators occurs when two three-level emitters enter into two-quantum resonances with other emitters of the second ensemble inverted relative dipole-forbidden transition. Similar collectivization processes can amplify (or inhibit) the collective spontaneous emission rate of each atomic sub-ensemble. The sign of exchange integral between the two atoms from different sub-ensembles depends on the retardation time and distance between them. This problem is connected to the possibilities of amplifying of entangled quanta and established the coherence between photon pairs. For this, the cooperative interaction of three-radiator subsystems is proposed in which one of them is inverted relative to the dipole-forbidden transitions, but another inverted dipole-active three-level system ignites this transition.

Taking into consideration the elementary acts of two-photon resonance between radiators, we have demonstrated the increasing of two-photon emission rate in one of the radiator subsystem comparison with traditional two-photon super-fluorescence [5]. The mutual influence of two- and single-photon super-fluorescent processes on the two-photon cooperative emission of the inverted subsystem relatively dipole-forbidden transition depends on the position of atoms in the exchange potential. Two possibilities of two- and three-particle exchanges through the vacuum field are represented in **Figure 1A–C**, taking into consideration the two-

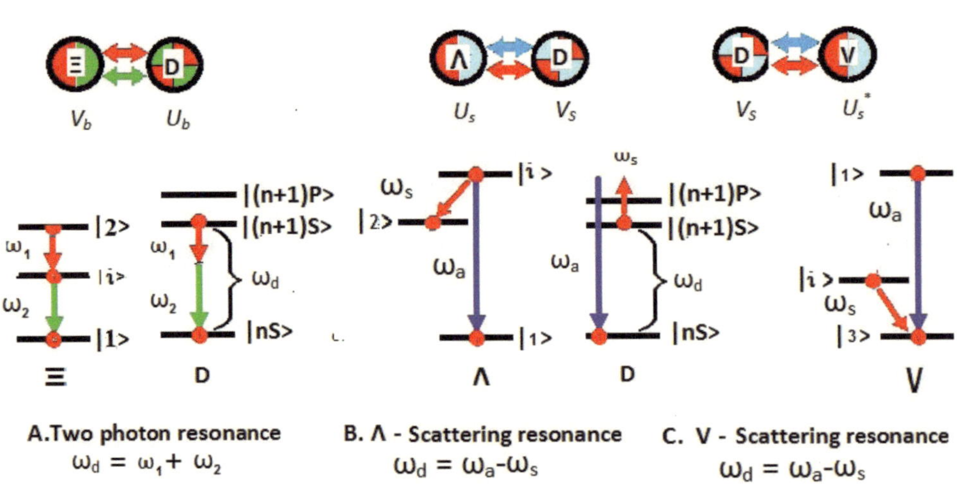

Figure 1.
The resonances between the two-photon transitions of D atomic subsystem and the three-level dipole-active systems in Ξ (A), V (B), and Λ (C) configurations. The three-level atoms are situated at relative distances $r_{d\xi}$, $r_{d\lambda}$, and r_{rv}. The exchange energies between the D subsystem in the two-photon resonance $\omega_0 = \omega_1 + \omega_2$ with the Ξ subsystem (A) and the scattering resonance $2\omega_0 = \omega_a - \omega_s$ with V subsystem (B) are given by the expressions (14) and (17).

photon resonance and scattering processes between the dipole-forbidden subsystem D and dipole-active subsystems of Ξ, Λ, and V, respectively. Here, the product of two vacuum polarizations of the atom Ξ (or V, Λ) comes into resonance with the polarization of the dipole-forbidden transitions of the D atom.

Using two small parameters in Section 2, we propose the projection operator method of elimination of the EMF operators from the generalized equation of atomic subsystems in single- and two-photon resonances. The possibilities of two-photon cooperative resonance between three-level radiators situated at a distance compared with the emission wavelength are demonstrated. Following this description the resonance interaction of a dipole-forbidden atom and three-level dipole-active radiator in the cascade configuration is described by the cooperative rate and the exchange integral (13). The similar expression (16) is obtained in the scattering process of three-level system in V or Λ—configurations with dipole-forbidden D subsystem represented in **Figure 1**. In Section 3 the spontaneous emission for the two radiators in the cascade or scattering resonances is given without the de-correlation of the atomic correlation functions between them.

2. Master equation of cooperative exchange between three-level radiators in two-quantum exchanges

Let us consider the interaction of three-level subsystems of radiators in V and Ξ configuration with D dipole-forbidden two-level ensemble through the vacuum of EMF. The Ξ three-level subsystem in cascade configuration, prepared in excited state $|2_\xi\rangle$, can pass into the Dicke super-radiance regime [1] relatively the dipole-active transitions $2_\xi \to \iota_\xi \to 1_\xi$ at frequencies ω_2 and ω_1 (**Figure 1A**). According to **Figure 1A**, the excited D atom relatively the dipole-forbidden transition $2_d \to 1_d$ passes in the ground state $|1_d\rangle$ simultaneously generating two quanta under the influence of cooperative decay of the Ξ three-level subsystem. Two-photon transition of the D-atom takes place through the virtual levels represented by the notations $|3_d\rangle$ with opposite parity relative to the ground $|1_d\rangle$ and excited $|2_d\rangle$ states, respectively. This case corresponds to the situation when the emission frequencies of the dipole-active Ξ radiators and D dipole-forbidden radiators satisfy the resonance condition $\omega_1 + \omega_2 = 2\omega_0$. Here ω_1 and ω_2 are the transition frequencies of the Ξ dipole-active radiators in Ξ, and $\hbar\omega_d = 2\hbar\omega_0$ is the energy distance between the ground $|nS\rangle$ and excited $|(n+1)S\rangle$ states of the dipole-forbidden transitions of D radiator (see **Figure 1A**).

The similar cooperative emissions can be observed in the two-quantum resonance interactions between the V (or Λ) three-level radiator in two quanta scattering interactions and the dipole-forbidden transitions of D atoms through the vacuum field (see **Figure 1B,C**). In this situation, we consider that the dipole-active transitions of the three-level radiator in the V (or Λ) configuration satisfy the scattering condition $\omega_a - \omega_s = \omega_d$ in interaction with the D subsystem. As it is represented in **Figure 1B**, the cone of the transition energies of the V or Λ dipole-active three-level atoms must be larger than the dipole-forbidden transition $|(n+1)S\rangle - |nS\rangle$ of atoms D, so that two-photon resonance between the two dipole-active transitions of V atom enters in the exact scattering resonance, $\omega_a - \omega_s = \omega_d$, with D atom. This nonlinear transition increases with the decreasing of the detuning from resonance with virtual $|3_d\rangle$ states of the D two-level system.

The Hamiltonian of the system consists of the free and interaction parts $H = \hat{H}_0 + \hat{H}_I$. Here the free part of this Hamiltonian is represented through the atomic and field operators:

$$\hat{H}_0 = \sum_k \hbar\omega_k \hat{a}_k^\dagger \hat{a}_k + \hbar \sum_{m=1}^{N} \omega_d \hat{D}_{zm} - \sum_{\alpha=s,}^{2} \sum_{al=1}^{N_\lambda} \hbar\omega_\alpha \hat{\Lambda}_{\alpha l}^{\alpha} \\ + \sum_{\alpha=s,}^{2} \sum_{al=1}^{N_v} \hbar\omega_\alpha \hat{V}_{\alpha l}^{\alpha} + \sum_{\alpha=1,} \sum_{2j=1}^{N_\xi} \hbar(-1)^\alpha \omega_\alpha \hat{\Xi}_{\alpha j}^{\alpha}, \tag{1}$$

where N, N_ξ, N_λ, and N_v are the number of atoms in the D, Ξ, Λ, and V subsystems, respectively; the energies of first and second levels of the Ξ, Λ, and V three-level subsystems are measured from the third intermediate state $|\iota\rangle$. The operators $\Xi_{1,j}^1$, $\Xi_{\iota,j}^{\iota}$, and $\Xi_{2,j}^2$ describe the population of the ground, intermediary, and excited states of the Ξ atom. The population operators of two excited and ground states $\hat{V}_{2,j}^2$, $\hat{V}_{1,j}^1$, and $\hat{V}_{\iota,j}^{\iota}$ can be introduced for the three-level atom in V configuration too. The similar expressions for two ground and one excited state can be introduced for Λ three-level atomic configuration $\hat{\Lambda}_{2,j}^2$, $\hat{\Lambda}_{1,j}^1$, and $\hat{\Lambda}_{\iota,j}^{\iota}$, respectively. The D atoms are considered as a two-level system, the state energy positions of which are measured from the middle point between the excited and ground states, respectively, $D_{z,j} = \left(D_{2,j}^2 - D_{1,j}^1\right)/2$. The first term of the Hamiltonian describes the free energy of EMF, the $k \equiv \mathrm{k}, \lambda$ modes of which is initially considered in the vacuum state $|0_k\rangle$. Here $\hat{a}_k$ and $\hat{a}_k^\dagger$ are annihilation and creation operators of EMF photons with wave vector k, polarization ε_λ, and the frequency ω_k, which satisfy the commutation relation $\left[\hat{a}_k^\dagger, \hat{a}_{k'}^\dagger\right] = \delta_{k,k'}$.

Taking into consideration the conservation energy laws, $\hbar(\omega_1 + \omega_2) = 2\hbar\omega_0$ and $\hbar(\omega_a - \omega_s) = 2\hbar\omega_0$ (according to **Figure 1A–C**, respectively), we introduce the interaction Hamiltonian $\hat{H}_I = \hat{H}_{I1} + \hat{H}_{I2}$ of the Ξ, Λ, V, and D subsystems with free EMF. Here $\hat{H}_{I1}$ describes the single-photon interaction of three-level atoms in the Ξ, V, and Λ configurations with a vacuum of EMF:

$$\hat{H}_{I1} = -\sum_k \sum_{j=1}^{N_\xi} \left[(\mu_{1\iota}, \mathrm{g}_k)\hat{\Xi}_{1j}^{\iota} + (\mu_{2\iota}, \mathrm{g}_k)\hat{\Xi}_{\iota j}^{2}\right] \hat{a}_k \exp\left[i(\mathrm{k}, \mathrm{r}_j)\right] \\ - \sum_k \sum_{l=1}^{N_\lambda} \left[(\mu_{\iota 1}, \mathrm{g}_k)\hat{\Lambda}_{1l}^{\iota} + (\mu_{\iota 2}, \mathrm{g}_k)\hat{\Lambda}_{2l}^{\iota}\right] \hat{a}_k \exp\left[i(\mathrm{k}, \mathrm{r}_l)\right] \\ - \sum_k \sum_{l=1}^{N_v} \left[(\mu_{\iota 1}, \mathrm{g}_k)\hat{V}_{\iota l}^{1} + (\mu_{\iota 2}, \mathrm{g}_k)\hat{V}_{\iota l}^{2}\right] \hat{a}_k \exp\left[i(\mathrm{k}, \mathrm{r}_l)\right] + H.c., \tag{2}$$

where $\varepsilon_1 \hat{H}_{I1}^{\Xi_1-} \sim \hat{\Xi}_{1j}^{\iota} \hat{a}_k$ and $\varepsilon_1 \hat{H}_{I1}^{\Xi_2=} \sim \hat{\Xi}_{\iota j}^{2} \hat{a}_k$ represent the two-photon cascade excitation of Ξ atom through the intermediary state $|\iota\rangle$; $\varepsilon_1 \hat{H}_{I1}^{S-} \sim \hat{\Lambda}_{2,j}^{\iota} \hat{a}_k$ (or $\varepsilon_1 \hat{H}_{I1}^{S-} \sim \hat{V}_{\iota,j}^{2} \hat{a}_k$) and $\varepsilon_1 \hat{H}_{I1}^{A-} \sim \hat{\Lambda}_{1,j}^{\iota} \hat{a}_k$ ($\varepsilon_1 \hat{H}_{I1}^{A-} \sim \hat{V}_{\iota,j}^{1} \hat{a}_k$) describe the excitation of Λ (or V) atom with the absorption of the photons with the energies $\hbar\omega_s$ and $\hbar\omega_a$, respectively. $\mu_{i,j}$ is dipole momentum transitions between the i and j states of the atoms. The second part of interaction Hamiltonian, $\hat{H}_{I2}$, describes the nonlinear interaction of the dipole-forbidden transition of D two-level system with vacuum field:

$$\hat{H}_{I2} = \sum_{k_1, k_2} \sum_{m=1}^{N} \left[q_s(\mathrm{k}_1, \mathrm{k}_2)\hat{D}_m^{-} \hat{a}_{k_2}^\dagger \hat{a}_{k_1} (1 - \delta_{k_1, k_2}) \exp\left[i(\mathrm{k}_1 - \mathrm{k}_2, \mathrm{r}_m)\right]\right. \\ \left. - q_b(k_1, k_2)\hat{D}_m^{+} \hat{a}_{k_2} \hat{a}_{k_1} \exp\left[i(\mathrm{k}_1 + \mathrm{k}_2, \mathrm{r}_m)\right]\right] + H.c. \tag{3}$$

This interaction is expressed by two-photon emission terms $\varepsilon_2 \hat{H}_{I2}^{b+} \sim \hat{D}_m^- \hat{a}_{k_2}^\dagger \hat{a}_{k_1}^\dagger$ and possible scattering of an emitted photon by the Ξ and V subsystems $\varepsilon_2 \breve{H}_{I2}s\pm \sim \hat{D}_m^\mp \hat{a}_{k_2}^\dagger \hat{a}_{k_1}$. The excitation and lowering operators of V, Λ, and Ξ dipole-active three-level subsystems are described by the operators of $U(3)$ algebra, which satisfy the commutation relations $\left[\hat{U}_{\beta j}^{\alpha}, \hat{U}_{\alpha' l}^{\beta'}\right] = \delta_{l,j}\left\{\hat{U}_{\alpha' j}^{\alpha}\delta_{\beta,\beta'} + \hat{U}_{\beta' j}^{\beta}\delta_{\alpha,\alpha'}\right\}$. Here the operator $\hat{U}_{\beta j}^{\alpha}$ is equivalent with V and Ξ operators, $\hat{V}_{\beta j}^{\alpha}$ and $\hat{\Xi}_{\beta j}^{\alpha}$, respectively. The inversion $\hat{D}_{lz}$ together with lowering and exciting $\hat{D}_j^{\pm}$ operators of D subsystem belongs to $SU(2)$ algebra: $\left[\hat{D}_{lz}, \hat{D}_j^{\pm}\right] = \pm\hat{D}_l^{\pm}\delta_{l,j}$ and $\left[\hat{D}_l^{+}, \hat{D}_m^{+}\right] = 2\delta_{l,m}\hat{D}_{lz}$. In comparison with single-photon interaction of Ξ and V atoms with vacuum field $\left(\mu_{i,j}, g_k\right)$, the nonlinear interaction of D two-level subsystem with EMF in two-photon and scattering interaction is described by the interaction constants and second order:

$$q_b(k_1, k_2) = \frac{\left(d_{31}, g_{k_1}\right)\left(d_{32}, g_{k_2}\right)}{2\hbar(\omega_{32} + \omega_{k_1})} + \frac{\left(d_{31}, g_{k_2}\right)\left(d_{32}, g_{k_1}\right)}{2\hbar(\omega_{31} - \omega_{k_2})},$$

$$q_s(k_1, k_2) = \frac{\left(d_{31}, g_{k_1}\right)\left(d_{32}, g_{k_2}\right)}{\hbar(\omega_{32} - \omega_{k_1})} + \frac{\left(d_{31}, g_{k_2}\right)\left(d_{32}, g_{k_1}\right)}{\hbar(\omega_{31} + \omega_{k_1})},$$

where $g_k = \varepsilon_\lambda\sqrt{2\pi\hbar\omega_k/V}$ and $d_{i,j}$ is dipole momentum transitions between the levels of the D atom. In the definition of the interaction parts of the Hamiltonian (2) and (3), we introduced the fictive small parameters ε_1 and ε_2 which will help us to establish the contributions of the second and third orders in two-photon decay rates.

In this section the conditions for which the pure super-fluorescence of the small number of radiators [14, 15] in the subsystems Ξ, V, and D enters into interaction during the delay time of cooperative spontaneous emission of each subsystem are considered, so that inhomogeneous broadening of excited atomic states can be neglected, $\tau_i \ll T_{2,i}$. Here $\tau_i = \tau_0/N_i$ is the collective time for which the polarization of the i subsystem becomes macroscopic; $T_{2,i}$ is the de-phasing time of the subsystem i, which includes the reciprocal inhomogeneous and Doppler-broadened line-width, $i \equiv \Xi$, V, and D (see, e.g., the papers [15, 16]). These conditions can be achieved using laser cooling method [17, 18] for three atomic ensembles represented in **Figure 1**A,B. Let us suppose that delay time of the super-radiant pulse is less than $T_{2,i}$; we will drop the terms connected with de-phasing time $T_{2,i}$ from the kinetic equations. In order to estimate the three-particle cooperative interaction, we will examine the situation in which one- and two-quantum interactions with the EMF bath are taken into account simultaneously. In this case it is necessary to eliminate from the density matrix equation the boson operators of EMF in nonlinear interaction with atomic subsystem. In comparison with the paper [12], here we will take into consideration the two-quantum effects connected with the influence of three-level atomic systems V and Ξ on the two-photon spontaneous emission of dipole-forbidden D subsystem. In this case instead of two dipole-active atoms, we can take into consideration only one three-level atom in two-photon resonance with dipole-forbidden system.

Let $\mathcal{P}$ be the projection operator for the complete density matrix $\breve{\rho}(t)$ on the vector basis of a free EMF subsystem $\rho_s(t) = \mathcal{P}\breve{\rho}(t)$ and $\breve{\rho}_b(t) = \overline{\mathcal{P}}\breve{\rho}(t)$, where $\breve{\rho}_s(t)$ and $\breve{\rho}_b(t)$ are slower and rapidly oscillating parts of the density matrix, respectively,

$\overline{\mathcal{P}} = 1 - \mathcal{P}$. It can be shown that $\mathcal{P}^2 = \mathcal{P}$ and $\overline{\mathcal{P}}\mathcal{P} = 0$. Recognizing that for $t = 0$ an electronic subsystem does not interact with the EMF, we define the projection operator $\mathcal{P} = \breve{\rho}_{ph}(0) \otimes Tr_{ph}\{\cdots\}$, where the trace is taking over the photon states and $\breve{\rho}_{ph}(0) = |0\rangle\langle 0|$ represents the density matrix of the vacuum of EMF. In this case one can represent the slow part of density matrix through the density matrix $\breve{W}(t) = Tr_{ph}\{\breve{\rho}(t)\}$ of the atomic subsystem $\breve{\rho}_s(t) = \breve{\rho}_{ph} \otimes \breve{W}(t)$, where $\breve{W}(0) = Tr_{ph}\breve{\rho}(0) = \breve{\rho}_r(0)$ is the density matrix of the prepared state of the atomic subsystem. The equations for the matrix $\breve{\rho}_s(t)$ and $\breve{\rho}_b(t)$ are

$$\frac{\partial \breve{\rho}_s(t)}{\partial t} = -i\mathcal{P}L_I(t)\{\breve{\rho}_s(t) + \breve{\rho}_b(t)\}, \tag{4}$$

$$\frac{\partial \breve{\rho}_b(t)}{\partial t} = -i\overline{\mathcal{P}}L_I(t)\{\breve{\rho}_s(t) + \breve{\rho}_b(t)\}, \tag{5}$$

where $\hat{L}_I(t) = \varepsilon_1 [\breve{H}_{I1}(t), \ldots]/\hbar + \varepsilon_2 [\breve{H}_{I2}(t), \ldots]/\hbar$ is the interaction part of Liouville operator. Following the known procedure of elimination of the rapidly oscillating part of the density matrix, we integrate Eq. (5) with respect to $\breve{\rho}_b(t)$ and substitute the resulting solution in Eq. (4). After this procedure we obtain the expression

$$\frac{\partial \breve{\rho}_s(t)}{\partial t} = -P \int_0^t d\tau L_I(t) U(t, t-\tau) L_I(t-\tau) \rho_s(t-\tau), \tag{6}$$

where the two-time evolution operator is represented by the T product $\breve{U}(t, t-\tau) = T\exp\left\{ -i\overline{\mathcal{P}} \int_{t-\tau}^{t} d\tau_1 L_I(\tau) \right\}$. In comparison with well-known procedure of the decomposition on the small parameter ε of the right-hand site of expression (6), here we have two parameters ε_1 and ε_2. The quantum correlation between the single- and two-photon interactions of atoms through the vacuum of the EMF can be found in the third order of the expansion on the small parameter product $\varepsilon_1^2 \varepsilon_2$ of the right-hand side of Eq. (6). Indeed considering the second and third order of the expansion on the small parameters ε_1 and ε_2, we represent the evolution operators $\breve{U}(t, t-\tau)$ and $\breve{\rho}_s(t-\tau)$ in the following approximate form $\breve{U}(t, t-\tau) \approx 1 - i\overline{\mathcal{P}} \int_{t-\tau}^{t} d\tau_1 L_i(\tau_1)$ and $\breve{\rho}_s(t-\tau) = \breve{\rho}_s(t) + \mathcal{P} \int_0^{\tau} d\tau_1 \hat{L}_i(t-\tau_1) \int_0^{t-\tau_1} d\tau_2 \hat{L}_i(t-\tau_1-\tau_2) \breve{\rho}_s(t-\tau_1-\tau_2)$. Upon substitution of this expression in Eq. (6), in the third order of small parameter λ, the equation for $\rho_s(t)$ becomes

$$\frac{\partial}{\partial t}\breve{\rho}_s(t) = -\mathcal{P} \int_0^t d\tau_1 \hat{L}_i(t) \left\{ \hat{L}_i(t-\tau_1) - i \int_{t-\tau_1}^{t} d\tau_2 \hat{L}_i(\tau_2) \right\} \hat{L}_i(t-\tau_1) \breve{\rho}_s(t). \tag{7}$$

Representing the Liouville operator, $\hat{L}_I(t)$, through single-, $L_{I1}(t) = \varepsilon_1 [\breve{H}_{I1}(t), \ldots]/\hbar$, and two-photon, $\lambda L_{I2}(t) = \varepsilon_2 [\breve{H}_{I2}(t), \ldots]/\hbar$, interaction parts, we can observe that in the third order on the decomposition on interaction Hamiltonian, the main contribution to the right-hand site of Eq. (7) gives the terms proportional to the $\varepsilon_1^2 \varepsilon_2$. Indeed, taking into consideration that the trace of an odd number of boson operator is zero, $Tr_{ph}\left\{ \rho_0 \breve{a}_{k_1}^{\dagger} \breve{a}_{k_2} {\dagger} \breve{a}_{k_3} \breve{a}_{k_4} \breve{a}_{k_5} \right\} = 0$, it is not difficult to observe that the projection of the operator product $\varepsilon_2^2\varepsilon_1\mathcal{P}\breve{H}_{I1}\breve{H}_{I2}\breve{H}_{I2}$ takes the zero value too. In the third order of the small parameters ε_i, the contribution of Liouville operator $\hat{L}_{I1}$ and $\hat{L}_{I2}$ must be found from the terms like $\mathcal{P}\hat{L}_{I1}\hat{L}_{I2}\hat{L}_{I1}\hat{\breve{\rho}}_s(t)$, which corresponds to two-photon resonances between the single- and two-photon transitions in the three-level atomic systems described by the Hamiltonian part (2) and (3), respectively. It is not difficult to observe that second-order decomposition on the interaction Hamiltonian gives zero contributions in the correlations between the Ξ, V, and D subsystems. This follows from the zero value of the trace of the odd number of boson operators, $Tr_{ph}\left\{\breve{\rho}_0\breve{a}^\dagger_{k_1}\breve{a}_{k_2}\breve{a}_{k_3}{}^\dagger\right\} = 0$, which corresponds to the projection of the operator product $\mathcal{P}\breve{H}_{I1}\breve{H}_{I2}\mathcal{P} = 0$.

Following this procedure of calculation of mean value of boson operators, it is observed that the two-photon resonance represented in **Figure 1A** can be described by the following diagrams:

$$\begin{aligned}\Delta\rho_3^b = i\lambda^3\int_0^t d\tau_1\int_0^{\tau_1} d\tau_2\Big\{&\mathcal{P}\hat{L}_{I1}^{\Xi_1-}(t)\hat{\hat{L}}_{I1}^{\Xi_2-}(t-\tau_2)\hat{L}_{I2}^{b+}(t-\tau_1)\rho_s(t)\\ &+\mathcal{P}\hat{L}_{I1}^{R-}(t)\hat{L}_{I1}^{S-}(t-\tau_2)\hat{L}_{I2}^{b+}(t-\tau_1)\rho_s(t)\\ &+\mathcal{P}\hat{L}_{I1}^{\Xi_2-}(t)\hat{L}_{I2}^{b+}(t-\tau_2)\hat{L}_{I1}^{\Xi_1-}(t-\tau_1)\rho_s(t)\\ &+\mathcal{P}\hat{L}_{I1}^{\Xi_1-}(t)\hat{L}_{I2}^{b+}(t-\tau_2)\hat{L}_{I1}^{\Xi_2-}(t-\tau_1)\rho_s(t)\\ &+\mathcal{P}\hat{L}_{I2}^{b+}(t)\hat{L}_{I1}^{\Xi_1-}(t-\tau_2)\hat{L}_{I1}^{\Xi_2-}(t-\tau_1)\rho_s(t)\\ &+\mathcal{P}\hat{L}_{I2}^{b+}(t)\hat{L}_{I1}^{\Xi_2-}(t-\tau_2)\hat{L}_{I1}^{\Xi_1-}(t-\tau_1)\rho_s(t)\Big\}+H.c.\end{aligned}\tag{8}$$

Here $L_{I1}^{\Xi_1-}(t) = \varepsilon_1\left[\breve{H}_{I1}^{\Xi_1-}(t),\dots\right]/\hbar$, $L_{I1}^{\Xi_2-}(t) = \varepsilon_1\left[\breve{H}_{I2}^{\Xi_2-}(t),\dots\right]/\hbar$, and $L_{I2}^{b-}(t) = \varepsilon_2\left[\breve{H}_{I2}^{b-}(t),\dots\right]/\hbar$ represent the Liouville operators of the interaction part of the Ξ and D atoms expressed through EMF annihilation and atomic exciting operators in the single- and two-quantum interactions.

The scattering resonance can be represented by the diagrams in which the conservation law $\omega_a - \omega_s = 2\omega_0$ must take place as represented in **Figure 1B**:

$$\begin{aligned}\Delta\rho_3^s = i\int_0^t d\tau_1\int_0^{\tau_1} d\tau_2\{&\mathcal{P}L_{I1}^{A-}(t)L_{I1}^{S+}(t-\tau_2)L_{I2}^{s+}(t-\tau_1)\rho_s(t)\\ &+\mathcal{P}L_{I1}^{S+}(t)L_{I1}^{A-}(t-\tau_2)L_{I2}^{s+}(t-\tau_1)\rho_s(t)\\ &+\mathcal{P}L_{I1}^{A-}(t)L_{I2}^{s+}(t-\tau_1)L_{I1}^{S+}(t-\tau_2)\rho_s(t)\\ &+\mathcal{P}L_{I1}^{S+}(t)L_{I2}^{s+}(t-\tau_1)L_{I1}^{A-}(t-\tau_2)\rho_s(t)\\ &+\mathcal{P}L_{I2}^{s+}(t)L_{I1}^{A-}(t-\tau_1)L_{I1}^{S+}(t-\tau_2)\rho_s(t)\\ &+\mathcal{P}L_{I2}^{s+}(t)L_{I1}^{S+}(t-\tau_1)L_{I1}^{A-}(t-\tau_2)\rho_s(t)\}+H.c.,\end{aligned}\tag{9}$$

where $L_{I2}^{s-}(t) = \varepsilon_2\left[\breve{H}_{I2}^{s-}(t),\dots\right]/\hbar$ is the Liouville parts for two-photon scarpering process of D atomic subsystem and $L_{I1}^{S-}(t) = \varepsilon_1\left[\breve{H}_{I1}^{S-}(t),\dots\right]/\hbar$ and

$L_{I1}^{A-}(t) = \varepsilon_1 \left[\breve{H}_{I1}^{A-}(t), \ldots \right] / \hbar$ correspond to the single-photon transitions in Ξ atomic subsystem described by the Hamiltonian parts (3) and (2), respectively.

So that after the trace on the EMF variables, we obtain $Tr\left\{ \hat{\rho}_{ph} \breve{a}_{k_1} \breve{a}_{k_3} \breve{a}_{k_2} \dagger \breve{a}_{k_4} \dagger \right\} = \left[\delta_{k_1 k_2} \delta_{k_3, k_4} + \delta_{k_1, k_4} \delta_{k_3, k_2} \right]$, $Tr\left\{ \hat{\rho}_{ph} \breve{a}_{k_1} \breve{a}_{k_2} \dagger \right\} = \delta_{k_1 k_2}$, and $Tr\left\{ \rho_{ph} \breve{a}_{k_2}^{\dagger} \breve{a}_{k_4} \dagger \breve{a}_{k_1} \breve{a}_{k_3} \right\} = 0$. We found the correlations between Ξ, V, and D atomic subsystem represented in **Figure 1**.

We found the correlations between Ξ, V, and D atomic subsystem represented in the **Figure 1**. Following projection technique procedures developed in Refs. [5, 13, 19], we find the terms of in the right-hand side of the master equation (7)–(9) for three species of radiators in interaction

$$\frac{d\breve{W}(t)}{dt} = \frac{d\breve{W}_0(t)}{dt} + \frac{d\breve{W}_{21b}(t)}{dt} + \frac{d\breve{W}_{21s}(t)}{dt}. \tag{10}$$

First term describes the cooperative single- and two-photon effects in each subsystem, respectively. Second term describes the exchanges between the single-photon processes of Ξ three-level subsystem and the two-photon transitions of the D radiators as this is represented in **Figure 1A**. The third term describes the scattering effect of the two radiators represented in **Figure 1B**.

All parameters and collective exchange integrals between the three-level radiators in V configuration and dipole-forbidden two-level system D are defined in the literature [1–12]:

$$\begin{aligned} \frac{d\breve{W}_0(t)}{dt} &= \frac{1}{2\tau_{\imath,1}} \sum_{l,j=1}^{N_\xi} \chi_1(j,l) \left[\breve{\Xi}_{\imath,j}^{1}, \breve{W}(t) \breve{\Xi}_{1,l}^{\imath} \right] + \frac{1}{2\tau_{\imath,2}} \sum_{l,j=1}^{N_\xi} \chi_2(j,l) \left[\breve{\Xi}_{2,j}^{\imath}, \breve{W}(t) \breve{\Xi}_{\imath,l}^{2} \right] \\ &+ \frac{1}{2\tau_{\imath,a}} \sum_{l,j=1}^{N_v} \chi_a(j,l) \left[\breve{V}_{1,j}^{\imath}, \breve{W}(t) \breve{V}_{\imath,l}^{1} \right] + \frac{1}{2\tau_{\imath,s}} \sum_{l,j=1}^{N_v} \chi_s(j,l) \left[\breve{V}_{2,j}^{\imath}, \breve{W}(t) \breve{V}_{\imath,l}^{2} \right] \\ &+ \frac{1}{2\tau_{\imath,s}} \sum_{l,j=1}^{N_\lambda} \chi_s(j,l) \left[\breve{\Lambda}_{\imath,j}^{2}, \breve{W}(t) \breve{\Lambda}_{2,l}^{\imath} \right] + \frac{1}{2\tau_{\imath,a}} \sum_{l,j=1}^{N_\lambda} \chi_a(j,l) \left[\breve{\Lambda}_{\imath,j}^{2}, \breve{W}(t) \breve{\Lambda}_{2,l}^{\imath} \right] \\ &+ \frac{1}{2\tau_d} \sum_{l,j=1}^{N} \chi_d(j,l) \left[\breve{D}_j^{-}, \breve{W}(t) \breve{D}_l + \right] + H.c., \end{aligned} \tag{11}$$

where $\tau_{\imath,\alpha} = 3\hbar c^3 / \left(4\mu_{\alpha,\imath}^2 \omega_\alpha^3 \right)$ is the spontaneous emission time of the dipole-active transitions $|\alpha\rangle \to |\imath\rangle$ of three-level atom in Ξ and V configurations and $\tau_d = \pi 3^2 \hbar^2 c^6 / \left(4^2 \omega_0^7 d_{23}^2 d_{31}^2 q_b^2(\omega_0, \omega_0) \right)$ is the two-photon spontaneous emission rate in the D atomic subsystem. This equation can be used for the description of interaction between the dipole-forbidden and dipole-active subsystems of radiators. For comparison of the real parts of the single- and two-photon exchange integrals, we can observe that the second decreases inversely proportional to the square distance r_{jl} between two D radiators: $\mathrm{Re}[\chi_a(j,l)] = \sin(\omega_a r_{j,l}/c) / [\omega_a r_{j,l}/c]$ and $\mathrm{Re}[\chi_d(j,l)] \sim \sin^2(\omega_0 r_{j,l}/c) / [\omega_0 r_{j,l}/c]^2$.

Following the two-parameter approach projection technique proposed in Ref. [13], $H_I 1 \sim \varepsilon_1$ and $H_I 2 \sim \varepsilon_2$, we easily found the three-particle exchanges between the radiators represented in **Figure 1A** described by master equation

$$\begin{aligned}\frac{d\breve{W}_{21b}(t)}{dt} = & -\frac{i}{4\tau_{12d}^{b}}\sum_{m=1}^{N}\sum_{l=1}^{N_\xi}\sum_{j=1}^{N_\xi}\{U_b(m,l)\\ & \times\left\{\left[\breve{D}_m^{-},\breve{W}(t)\breve{\Xi}_{\iota,l}^{2}\breve{\Xi}_{1,j}^{\iota}\right]+\left[\breve{D}_m^{-},\breve{W}(t)\breve{\Xi}_{1,j}^{\iota}\breve{\Xi}_{\iota,l}^{2}\right]\right\}\\ & + U_b^{*}(j,l,m)\left[\breve{\Xi}_{\iota,l}^{2},\left[\breve{\Xi}_{1,j}^{\iota},\breve{D}_m-\breve{W}(t)\right]\right]\\ & + [\breve{\Xi}_{1,j}^{\iota},[\breve{\Xi}_{\iota,l}^{2},\breve{D}_m-\breve{W}(t)]]\Big\} - \frac{i}{2\tau_{12d}^{b}}\sum_{m=1}^{N}\sum_{j=1}^{N_\xi}\sum_{l=1}^{N_\xi}V_b(j,m,l)\\ & \times\left\{\left[\breve{D}_m^{-}\breve{W}(t)\breve{\Xi}_{\iota l}^{2},\breve{\Xi}_{1,j}^{\iota}\right]+\left[\breve{D}_m^{-}\breve{W}(t)\breve{\Xi}_{1,j}^{\iota},\breve{\Xi}_{\iota,l}^{2}\right]\right\}\\ & + H.c.\end{aligned} \tag{12}$$

Here for $\omega_s \simeq \omega_r$, we have found the following integrals:

$$\begin{aligned}\frac{1}{\tau_{12d}^{b}} &= \left\{\frac{4}{3}\right\}^2 \frac{\omega_s^3(\omega_r)^3\mu_{\iota 2}\mu_{\iota 1}d_{23}d_{31}}{2\hbar^2c^6}\left[\frac{1}{\omega_{32}+\omega_2}+\frac{1}{\omega_{31}+\omega_1}\right],\\ V_b(j,m,l) &\simeq \frac{c^2[\exp[-i\omega_2 r_{ml}/c]-1][\exp[i\omega_1 r_{jm}/c]-1]}{\omega_1\omega_2 r_{jm}r_{ml}},\\ U_b(j,m,l) &= \exp[-i\omega_1 r_{mj}/c]V_b(j,l,m).\end{aligned} \tag{13}$$

Here $1/\tau_{12d}^{b}$ is the three-particle cooperative emission rate of two atoms from Ξ subsystems and one atom from D ensemble situated at the relatively small distance $r_{jl} \ll \lambda_{s(r)}$. $V_b(j,m,l)$ is the exchange integral which describes the influence of the m atom from D ensemble on the single-photon transitions of the j and l radiators from the Ξ subsystem. $U_b(j,m,l)$ is the inverse process of the cooperative action of j and l radiators from the Ξ ensemble on the two-photon transitions of m radiator from the D subsystem.

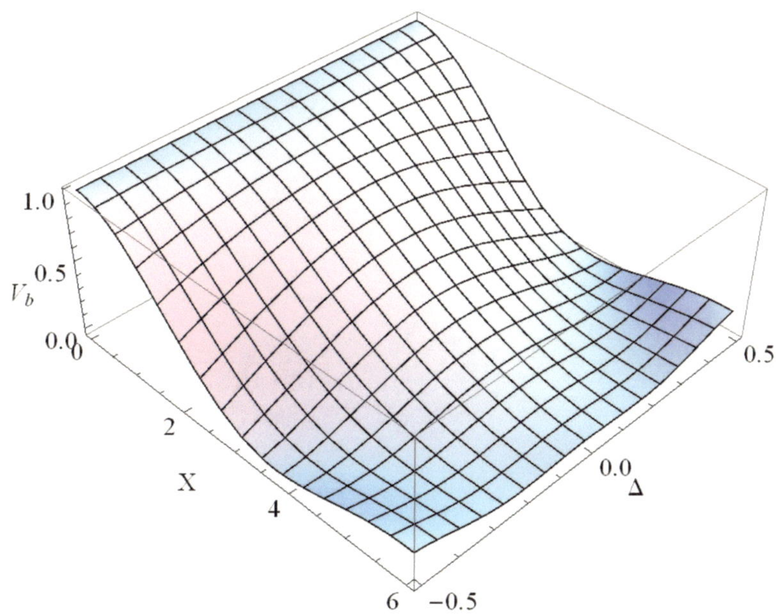

Figure 2.
The real part of exchange integral V_b, defined in expression (14), is plotted as a function of relative distance between radiators, $X = \omega_0 r/c$, and relative displacement, $\Delta = (\omega_1 - \omega_0)/\omega_0$.

For two atoms represented in **Figure 1A**, the simple exchange integral between these radiators can be obtained from expression (13):

$$V_b = \frac{\lambda_1\lambda_2[\exp(-2i\pi r/\lambda_2) - 1][\exp(2i\pi r/\lambda_1) - 1]}{(2\pi r)^2}, \tag{14}$$

where λ_2 and λ_1 are the emission wavelengths in cascade transition of the dipole-active three radiators in Ξ configuration, situated at distance r. The real part of this function describes the three-particle decay rate of the system. The dependence of exchange integral (14) on the relative distance between the Ξ and D atoms (14), $X = \omega_0 r/c$ and the displacement, $\Delta = (\omega_1 - \omega_0)/\omega_0$ relatively the degenerate frequency ω_0, is plotted in **Figure 2**. As follows from this dependence, the exchange integral achieved the maximal radius, when $\omega_1 = \omega_2$, which corresponds to the situation $\Delta = 0$.

The part of master Equation (10) for resonance scattering interaction between the absorbed and emitted photons by the dipole-active Λ and V subsystems and D dipole-forbidden radiators can be obtained from the third-order expansion on the smallest parameter λ. In this situation, the scattering part of the master equation represented by the scheme 1 B becomes

$$\begin{aligned}\frac{d\breve{W}_{21s}(t)}{dt} = {} & \frac{i}{2\tau^{s}_{sad}} \sum_{m,j,l=1} \Big\{U_s(j,m,l)[\breve{V}^{1}_{\iota,j}, \breve{V}^{l}_{2,l}\breve{D}_m - \hat{W}(t)] \\ & + U^{*}_{s}(j,m,l)[\breve{V}^{l}_{2,l}, \breve{W}(t)\breve{V}^{1}_{\iota,j}\breve{D}_m -]\Big\} \\ & - \frac{i}{2\tau^{s}_{sad}} \sum_{m,j,l=1} V_s(j,m,l)\left[\breve{D}^{-}_{m}, \breve{V}^{l}_{2,l}\breve{W}(t)\breve{V}^{1}_{\iota,j}\right] \\ & + H.c.,\end{aligned} \tag{15}$$

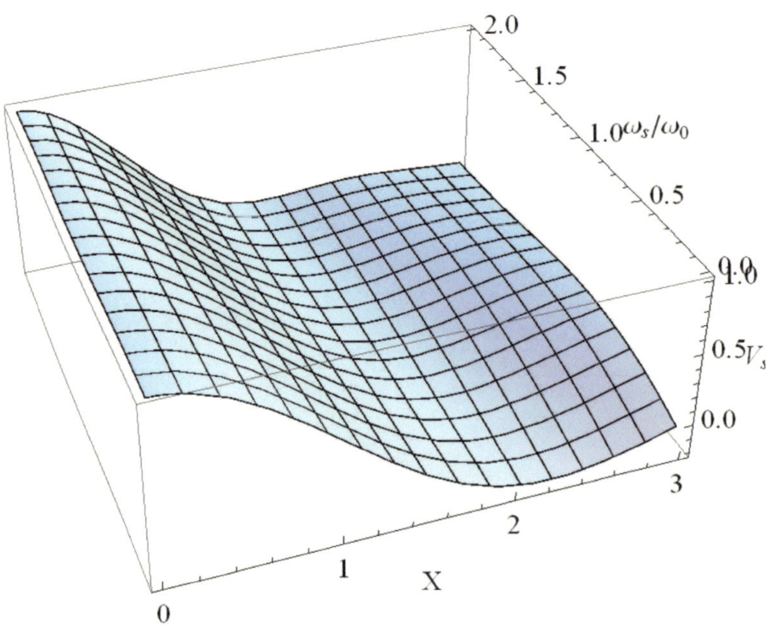

Figure 3.
The real part of the scattering exchange integrals V_s, defined in expressions (17), is plotted as a function of relative distance between radiators, $X = \omega_0 r/c$, and relative scattering frequency, ω_s/ω_0.

where

$$\frac{1}{\tau_{sad}^{s}} = \left(\frac{4}{3}\right)^{2} \frac{\mu_{\iota 2}\mu_{\iota 1}d_{23}d_{31}\omega_s^3(\omega_a)^3}{c^6\hbar^2}\left[\frac{1}{\omega_{32}-\omega_s}+\frac{1}{\omega_{31}+\omega_s}\right],$$

$$V_s(j,m.l) = \frac{(\exp[-i\omega_s r_{ml}/c]-1)(\exp[i\omega_a r_{mj}/c]-1)}{\omega_s\omega_a(r_{ml}/c)(r_{mj}/c)}, \tag{16}$$

$$U_s(j,m,l) = \exp[-i\omega_s r_{ml}/c]V_s(j,m.l).$$

First term in Eq. (15) describes the transition of D atom under the influence of the scattering process of emitted photons of the atoms from V subsystem. This process is described by exchange integral $V_s(j,m.l)$. The last two terms in master Eq. (15) describe the scattering process of emitted photons by the V atoms under the influence of D subsystem.

The similar expression is obtained for the interaction of Λ three-level radiator with D atom represented in **Figure 1C**. In this case we must replace the operators of V subsystem in expression (15) with corresponding transition operators of Λ system $\breve{V}_{\iota,j}^{1} \to \hat{\Lambda}_{1j}^{\iota}$: $\breve{V}_{2,l}^{\iota} \to \breve{\Lambda}_{\iota,l}^{2}$ and their Hermit conjugated operators.

For the two atoms, expression (16) was reduced to the simple representation

$$V_s = \frac{\lambda_s\lambda_a[1-\exp(-2i\pi r/\lambda_s)][1-\exp(2i\pi r/\lambda_a)]}{(2\pi r)^2}. \tag{17}$$

Here the wavelength λ_s (λ_a) corresponds to the emitted photons at Stokes or anti-Stokes frequencies represented in **Figure 1**. The numerical representation of the real part of the exchange integral (17) as the function of the relive distance between the atoms $X = \omega_0 r/c$ and the relative Stokes frequency ω_s/ω_0 is plotted in the **Figure 3**. It is observing the nonsignificant dependence of this exchange integral on the frequency ω_s. The significant dependence on the detuning from resonance can be observed in the dependence of cooperative rate $1/\tau_{sad}^{s}$ represented in expressions (16).

In this section we obtained the correlations between dipole-active and dipole-forbidden subsystems of radiators, where the two-quantum exchange integral has the same magnitude as the two-photon quantum interaction between atoms of D subsystem. In the case of the big number of radiators in each subsystem, the correlated terms, expressions (12) and (15), give the cubic contribution in the cooperative diagrams of the kinetic equation $\varepsilon_1^2\varepsilon_2 NN_\xi^2$. When $N = N_\xi$ these terms can archived the value proportional to the Dicke super-radiance [1] even for the same small parameters of each subsystem $\varepsilon_1 = \varepsilon_2$. In this case the number of atoms in each subsystem must achieve the value for which the third order has the same magnitude as the second order $\varepsilon^2N^2 \sim \varepsilon^3N^3$. In conclusion we observe that the decomposition on the small parameter ε can be regarded as a sum of single- and the two-photon transition amplitudes proportional to ε_1 and ε_2, where $\varepsilon_1 \sim (\mu_{1\iota}, g_k)$ and $\varepsilon_2 \sim q_b(k_1,k_2)$ or $q_s(k_1,k_2)$. Considering the situation when the two-photon amplitude is smaller than the single-photon amplitude $\varepsilon_2 < \varepsilon_1$, we conclude that beginning with the third-order term, the correlation diagrams (12) and (15), proportional to $\varepsilon_1^2\varepsilon_2$, can play an important role in the two-quantum decay process even for the two-atomic system consisted from one atom of each subsystems: D and Ξ (or D and V). For example, in the situation when $\varepsilon_1 = 0.7$ and $\varepsilon_2 = 0.25$, the magnitude of two-photon emission, $\varepsilon_2^2 = 0.0625$, becomes smaller than the cooperative magnitude $\varepsilon_1^2\varepsilon_2 = 0.1225$). In other words we can find the condition for which we can neglect the decay rate of two-photon emission of the D atom in comparison with the cooperative effect described by expressions (12) and (15). This possibility to control the two-photon decay process of D atom with the decay process of Ξ or V excited three-level atom is given in the next section.

3. Two-photon energy transfer between the two three-level radiators

Master Eq. (10) can be used for the description of cooperative interaction between the dipole-forbidden and dipole-active radiators in two-photon exchanges. Indeed passing again from Schrodinger to Heisenberg pictures $Tr\left[\hat{W}(t)\hat{O}(0)\right] = Tr\left[\hat{W}(0)\hat{O}(t)\right]$, we can obtain from this expression the equation of the arbitrary atomic operator $\hat{O}(t)$. Let us firstly discuss the nonlinear interaction in which Ξ and D atoms enter in two-photon resonance as represented in **Figure 1A**. Studying the cooperative interaction between the dipole-forbidden and dipole-active radiators, the closed system of equations for the correlation functions can be found in such approach. Considering that the numbers of atoms in the each subsystem are relatively small, we can obtain the following generalized equation for the arbitrary operator O_b:

$$
\begin{aligned}
\frac{d\langle O_b(t)\rangle}{dt} &= \frac{1}{2\tau_{\iota,1}}\sum_{l,j=1}^{N_\xi}\chi_1(j,l)\left\langle\hat{\Xi}^{\iota}_{1,l}(t)\left[\hat{O}_b(t),\hat{\Xi}^{1}_{\iota,j}(t)\right]\right\rangle \\
&+\frac{1}{2\tau_{\iota,2}}\sum_{l,j=1}^{N_\xi}\chi_2(j,l)\left\langle\hat{\Xi}^{2}_{\iota,l}(t)\left[\hat{O}_b(t),\hat{\Xi}^{\iota}_{2,j}(t)\right]\right\rangle \\
&+\frac{1}{2\tau_d}\sum_{l,j=1}^{N}\chi_d(j,l)\left\langle\hat{D}^{+}_{l}(t)\left[\hat{O}_b(t),\hat{D}^{-}_{j}(t)\right]\right\rangle \\
&-\frac{i}{4\tau^{b}_{12d}}\sum_{m=1}^{N}\sum_{l=1}^{N_\xi}\sum_{j=1}^{N_\xi}\left\{U_b(m,l,j)[\left\langle\hat{\Xi}^{2}_{\iota,l}(t)\hat{\Xi}^{\iota}_{1,j}(t)\left[\hat{O}_b(t),\hat{D}^{-}_{m}(t)\right]\right\rangle\right. \\
&+\left\langle\hat{\Xi}^{\iota}_{1j}(t)\hat{\Xi}^{2}_{\iota l}(t)\left[\hat{O}_b(t),\hat{D}^{-}_{m}(t)\right]\right\rangle] + U^{*}_{b}(j,l,m)\langle\left\{\left[\left[\hat{O}_b(t),\hat{\Xi}^{\iota}_{1,j}(t)\right],\hat{\Xi}^{2}_{\iota,l}(t)\right]\right. \\
&\left.\left.+\left[\left[\hat{O}_b(t),\hat{\Xi}^{2}_{\iota,l}(t)\right],\hat{\Xi}^{\iota}_{1,j}(t)\right]\right\}\hat{D}^{-}_{m}\rangle\right\} \\
&-\frac{i}{2\tau^{b}_{12d}}\sum_{m=1}^{N}\sum_{j=1}^{N_\xi}\sum_{l=1}^{N_\xi}V_b(j,m,l)\left\{\langle\hat{\Xi}^{\iota}_{1,j}(t)[\hat{\Xi}^{2}_{\iota,l}(t),\hat{O}_b(t)]\hat{D}^{-}_{m}(t)\rangle\right. \\
&\left.+\langle\hat{\Xi}^{2}_{\iota,l}(t)[\hat{\Xi}^{\iota}_{1,j}(t),\hat{O}_b(t)]\hat{D}^{-}_{m}(t)\rangle\right\} + H.c.
\end{aligned}
\tag{18}
$$

In order to simplify this problem, we analyze below the situation in which we have only a single atom in each subsystem. In this case we can replace the operator O_b with the excitation numbers operators $\hat{N}_\alpha = \hat{\Xi}^{\alpha}_{\alpha}(t)$ and $\hat{N}_d = \hat{D}_z + 0.5$ of Ξ and D atoms, respectively. Here $\alpha = 1, 2$ and ι. When emission frequencies of the one-photon radiators coincide with $\omega_1 \simeq \omega_2 \simeq \omega_0$, the dependence (14) becomes real and positive defined function Ξ and D radiators. Here $[\exp[i\omega_0 r/c] - 1][\exp[-i\omega_0 r/c]$

$-1 = 2[1 - \cos(\omega_0 r/c)]$. According to this expression, the exchange integrals become

$$V^b_{12d} = \frac{2(1 - \cos(2\pi r/\lambda_0)}{(2\pi r/\lambda_0)^2}, \quad U^b_{12d} = \exp[-i\omega_0 r/c]V_{srd}.$$

In this case one can introduce the expression exchange rate $1/\tau^b_{srd}$ as a function of the distance between the dipole-active and dipole-forbidden subsystems:

$$\frac{1}{\tau^b_{12d}(r)} = \frac{1}{\tau^b_{srd}} \frac{2(1 - \cos(2\pi r/\lambda_s)}{(2\pi r/\lambda_s)^2}, \tag{19}$$

where $\lambda_0 = c/[2\pi\omega_0]$. Taking into account the above definitions and introducing the correlation functions between the polarizations of Ξ and D atoms $\langle \hat{F}_b(t,r) \rangle = i\left[\left\langle \hat{\Xi}^2_1(t)\hat{D}^-(t)\right\rangle - \left\langle \hat{D}^+(t)\hat{\Xi}^1_2(t)\right\rangle\right]$ and $\langle \hat{E}_b(t,x) \rangle = \left\langle \hat{\Xi}^2_1(t)\hat{D}^-(t)\right\rangle + \left\langle \hat{D}^+(t)\hat{\Xi}^1_2(t)\right\rangle$, we obtain the closed system of equations from expression (18):

$$\begin{aligned}
\frac{d}{dt}\langle \hat{N}_2(t,x)\rangle &= -\frac{\langle \hat{N}_2(t)\rangle}{\tau_{\iota,2}} - \frac{1}{4\tau^b_{12d}(x)}\left[\cos(x)\langle \hat{F}_b(t,x)\rangle - \sin(x)\langle \hat{E}(t,x)\rangle\right], \\
\frac{d}{dt}\langle \hat{N}_\iota(t,x)\rangle &= \frac{\langle \hat{N}_2(t,x)\rangle}{\tau_{\iota,2}} - \frac{\langle \hat{N}_\iota(t,x)\rangle}{\tau_{\iota,1}} + \frac{1}{2\tau^b_{12d}(x)}\left[\cos(x)\langle \hat{F}_b(t,x)\rangle\right. \\
&\quad \left. - \sin(x)\langle \hat{E}(t,x)\rangle\right] + \frac{\langle \hat{F}_b(t,x)\rangle}{2\tau^b_{12d}(x)}, \\
\frac{d}{dt}\langle \hat{N}_1(t,x)\rangle &= \frac{\langle \hat{N}_\iota(t)\rangle}{\tau_{\iota,1}} - \frac{1}{4\tau^b_{12d}(x)}\left[\cos(x)\langle \hat{F}_b(t,x)\rangle - \sin(x)\langle \hat{E}_b(t,x)\rangle\right] \\
&\quad - \frac{1}{2\tau^b_{12d}(x)}\langle \hat{F}_b(t,x)\rangle; \\
\frac{d}{dt}\langle \hat{N}_d(t,x)\rangle &= -\frac{\langle \hat{N}_d(t,x)\rangle}{\tau_d} \\
&\quad + \frac{1}{4\tau^b_{12d}(x)}\left[\cos(x)\langle \hat{F}_b(t,x)\rangle + \sin(x)\langle \hat{E}_b(t,x)\rangle\right], \\
\frac{d}{dt}\langle \hat{F}_b(t,x)\rangle &= -\frac{\langle \hat{F}_b(t,x)\rangle}{2}\left(\frac{1}{\tau_d} + \frac{1}{\tau_{\iota,2}}\right) \\
&\quad - \frac{1}{2\tau^b_{12d}(x)}\left\{\cos(x)\left[2\langle \hat{N}_2(t,x)\hat{N}_d(t)\rangle - \langle N_2(t,x)\rangle\right.\right. \\
&\quad - \langle \hat{N}_d(t)(\hat{N}_i(t) - \hat{N}_2(t))\rangle + \langle \hat{N}_d(t)(1 - \hat{N}_2(t) - 2\hat{N}_i(t)\rangle] \\
&\quad \left. - 2\langle \hat{N}_d(t)\hat{N}_i(t)\rangle + 2\langle \hat{N}_d(t)\hat{N}_2(t)\rangle\right\}; \\
\frac{d\langle \hat{E}_b(t,x)\rangle}{dt} &= \frac{\langle \hat{E}_b(t,x)\rangle}{2}\left(\frac{1}{\tau_d} + \frac{1}{\tau_{\iota,2}}\right) - \frac{1}{2\tau^b_{12d}(x)}\sin(x)\left[2\langle \hat{N}_2(t)\hat{N}_d(t)\rangle\right. \\
&\quad - \langle \hat{N}_2(t)\rangle + \langle \hat{N}_d(t)(\hat{N}_i(t) - \hat{N}_2(t))\rangle \\
&\quad \left. - \langle \hat{N}_d(t)(1 - \hat{N}_2(t) - 2\hat{N}_i(t))\rangle\right]; \\
\frac{d}{dt}\langle \hat{N}_2(t)\hat{N}_d(t)\rangle &= -\langle \hat{N}_2(t)\hat{N}_d(t)\rangle\left[\frac{1}{\tau_{\iota,2}} + \frac{1}{\tau_d}\right], \\
\frac{d}{dt}\langle \hat{N}_\iota(t)\hat{N}_d(t)\rangle &= \frac{\langle \hat{N}_2(t)\hat{N}_d(t)\rangle}{\tau_{\iota,2}} - \langle \hat{N}_\iota(t)\hat{N}_d(t)\rangle\left[\frac{1}{\tau_{\iota,1}} + \frac{1}{\tau_d}\right].
\end{aligned} \tag{20}$$

Using this system of Eq. (20), we can numerically study the cooperative nonlinear exchanges through the vacuum field between the Ξ and D radiators situated at relative distance x. One can observe that the spontaneous generation of photon pair by the D atom is drastically modified by the time increase of the cooperative correlation between the radiators. Indeed considering that the decay rate of the D atom $1/\tau_d$ is smaller than similar rates of the cascade transition in the Ξ atom ($\tau_d/\tau_{\xi,i} \simeq 6$; $\tau_d/(4\tau_{12d}) = 2$), we can numerically represent this dependence as a function of the relative time, t/τ_d, and the relative distance between the radiators, $x = 2\pi r/\lambda_0$. As shown in **Figure 4A**, the decay rate of D atom is drastically modified at small distances between the radiators which is in accordance with the analytic expressions (19). Considering that both atoms Ξ and D are prepared in the excited state, we observe the significant enhancement of the two-photon emission rate of the D radiator under the influence of the Ξ decay process.

Let us simplify the system of Eq. (20) in order to solve it exactly. Indeed, when dipole-active Ξ atom is situated at small distance relative to the D radiator ($x \ll 1$), the system of Eq. (20) is drastically simplified:

$$\begin{aligned}
\frac{d}{dt}\langle \hat{N}_2(t)\rangle &= -\frac{\langle \hat{N}_2(t)\rangle}{\tau_{\iota,2}} - \frac{\langle \hat{F}_b(t)\rangle}{4\tau_{12d}^b},\\
\frac{d}{dt}\langle \hat{N}_d(t)\rangle &= -\frac{\langle \hat{N}_d(t)\rangle}{\tau_d} + \frac{1}{4\tau_{12d}^b}\langle \hat{F}_b(t)\rangle,\\
\frac{d}{dt}\langle \hat{F}_b(t)\rangle &= -\frac{\langle \hat{F}_b(t)\rangle}{2}\left(\frac{1}{\tau_d}+\frac{1}{\tau_{\iota,2}}\right)\\
&\quad -\frac{1}{2\tau_{12d}^b}[4\langle \hat{N}_2(t)\hat{N}_d(t)\rangle + \langle \hat{N}_d(t)\rangle\\
&\quad -\langle \hat{N}_2(t)\rangle - 5\langle \hat{N}_d\hat{N}_\iota\rangle],\\
\frac{d}{dt}\langle \hat{N}_2(t)\hat{N}_d(t)\rangle &= -\langle \hat{N}_2(t)\hat{N}_d(t)\rangle\left[\frac{1}{\tau_{\iota,2}}+\frac{1}{\tau_d}\right],\\
\frac{d}{dt}\langle \hat{N}_\iota(t)\hat{N}_d(t)\rangle &= \frac{\langle \hat{N}_2(t)\hat{N}_d(t)\rangle}{\tau_{\iota,2}} - \langle \hat{N}_\iota(t)\hat{N}_d(t)\rangle\left[\frac{1}{\tau_{\iota,1}}+\frac{1}{\tau_d}\right].
\end{aligned} \tag{21}$$

The exact solution of this linear system of equation can be represented through solution of characteristic equation $Y_\alpha = \sum_{j=1}^{5} C_\alpha^j \exp\left[\Theta_j t\right]$, where $\alpha = 1, 2, 3, 4, 5$ and $\{Y_\alpha\}$ are the atomic functions, $Y_1(t) = \langle \hat{N}_d(t)\rangle$, $Y_2(t) = \langle \hat{N}_2(t)\rangle$, $Y_3(t) = \langle \hat{F}_b(t)\rangle$, $Y_4(t) = \langle\langle \hat{N}_2(t)\hat{N}_d(t)\rangle$, and $Y_5(t) = \langle\langle \hat{N}_\iota(t)\hat{N}_d(t)\rangle$; the solution of characteristic equation is

$$\begin{aligned}
&\Theta_1 = -\left(\frac{1}{\tau_2}+\frac{1}{\tau_d}\right);\quad \Theta_2 = -\left(\frac{1}{\tau_1}+\frac{1}{\tau_d}\right);\quad \Theta_3 = -\frac{1}{2}\left(\frac{1}{\tau_d}+\frac{1}{\tau_{\iota,2}}\right);\\
&\Theta_{4,5} = -\frac{1}{2}\left\{\frac{1}{\tau_{\iota,2}}+\frac{1}{\tau_d} \pm \sqrt{\left(\frac{1}{\tau_d}-\frac{1}{\tau_{\iota,2}}\right)^2 - \frac{1}{\tau_{12b}^2}}\right\}.
\end{aligned} \tag{22}$$

The coefficients $\{C_\alpha^j\}$ are determined from the initial conditions. As follows from the numerical estimation plotted in **Figure 4B** and solutions of characteristic

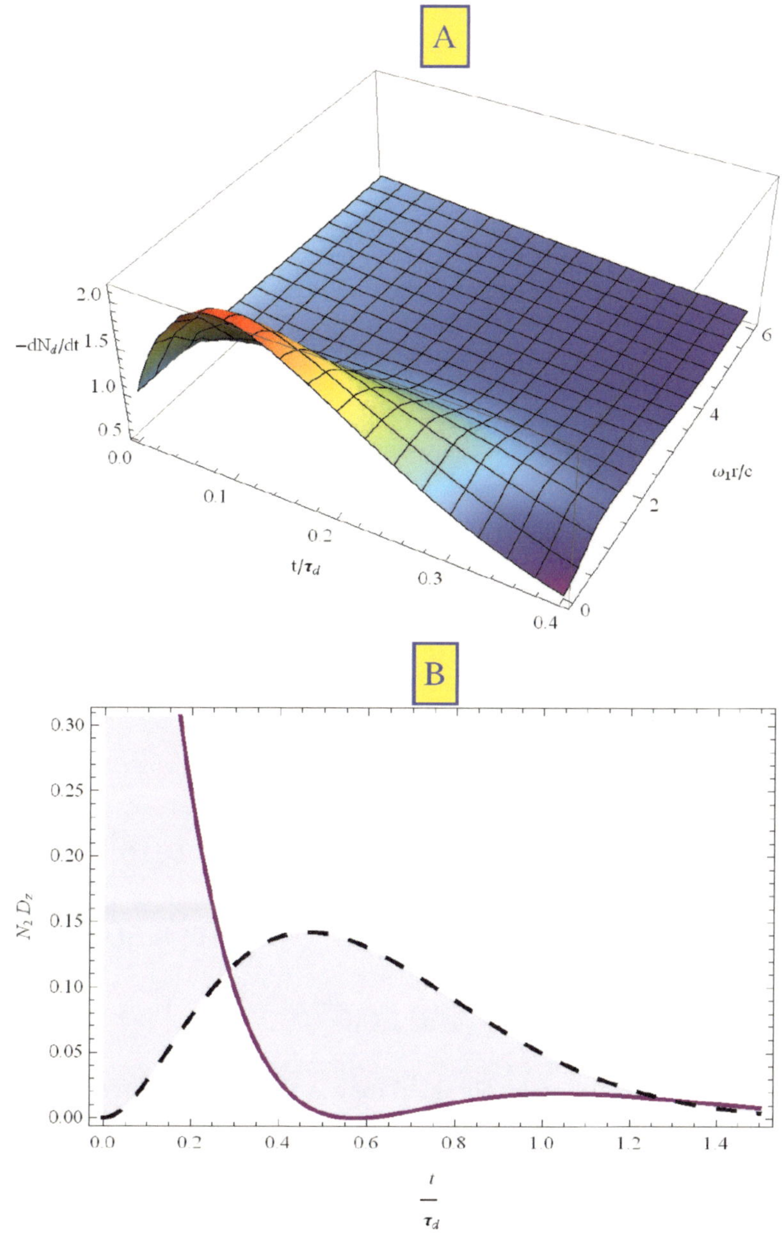

Figure 4.
(A) The decay rate $-d\langle N_d/dt\rangle$ of the dipole-forbidden transitions of the D radiator under the influence of Ξ three-level radiator. This solution of Eq. (21) is plotted as function of t/τ_d and relative distance between the radiators $x = \omega_1 r/c$, for the following parameters of the system: $\langle N_1\rangle = \langle N_\iota\rangle = 0.$, $\langle N_2\rangle = 1.$, $\langle N_d\rangle = 1$, $\tau_d/\tau_1 = \tau_d/\tau_2 = 6$, and $\tau_d/(4\tau_{12d}) = 2$. (B) The decay process of excited state $|2\rangle$ of three-level system (thick line) and the transfer of the excitation from the Ξ radiator to D atom (dashed line) in the process of cascade emission of Ξ atom situated at relative distance $x < < 1$ for the same parameters of the system and excitation conditions: $\langle N_1\rangle = \langle N_\iota\rangle = 0$, $\langle N_2\rangle = 1.$, and $\langle N_d\rangle = 0$.

in Eq. (22), the oscillatory decay of the atomic inversion is possible, when $1/\tau_d = 1/\tau_{\iota,2}$. In this case the solutions $\Theta_{4,5}$ become complex. We observe such an oscillation of the atoms inversion of Ξ radiator prepared initially in the excited state. In this process the rate of energy transfer from Ξ to D atoms represented in **Figure 4B** has the oscillator behavior. In the case of the excitation of D, the coupling between the radiators becomes more effective, when the virtual level of the D atom is situated between the excited and ground states (see **Figure 4B**). As the virtual states of the D radiator is off from the resonance with the dipole-active transitions of the Ξ radiators, the excitation of D atom takes place only with the absorption of both emitted photons by the Ξ atom. The cooperative effects between the Ξ and D radiators are described by second-order correlation function $G_2 = \left\langle \hat{E}^-(t)\hat{E}^-(t)\hat{E}^+(t)\hat{E}^+(t) \right\rangle = G_2^0 + \alpha\langle \hat{F}_b(t)\rangle$. Here G_2^0 was derived in Ref. [5]. The contribution to the second-order correlation function remains larger than the square value of the first-order correlation function $G_1 = \langle E^-(t)E^+(t)\rangle$, so that we can conclude that new cooperative effects between single- and two-photon transitions of D and Ξ subsystems play an important role in the two-photon decay process. Let us now return to the V three-level system in scattering interaction with the D system as this is represented in **Figure 1B**. In accordance with master Eq. (10) and its analytic representation (15), we can obtain the following expression for arbitrary atomic operators $\hat{O}_s(t)$.

$$
\begin{aligned}
\frac{d\left\langle \hat{O}_s(t) \right\rangle}{dt} &= \frac{1}{2\tau_{\iota,1}} \sum_{l,j=1}^{N_v} \chi_a(j,l)\left\langle \hat{V}^1_{\iota,l}(t)\left[\hat{O}^{(s)}(t), \hat{V}^{\iota}_{1,j}(t)\right]\right\rangle \\
&+ \frac{1}{2\tau_{\iota,2}} \sum_{l,j=1}^{N_v} \chi_s(j,l)\left\langle \hat{V}^2_{\iota,l}(t)\left[\hat{O}^{(s)}(t), \hat{V}^{\iota}_{2,j}(t)\right]\right\rangle \\
&+ \frac{1}{2\tau_d} \sum_{l,j=1}^{N} \chi_d(j,l)\left\langle \hat{D}^+_l(t)\left[\hat{O}^{(s)}(t), \hat{D}^-_j(t)\right]\right\rangle \\
&- \frac{i}{2\tau^s_{sad}} \sum_{m=1}^{N}\sum_{j=1}^{N_v}\sum_{l=1}^{N_v} V_s(j,m,l)\left\langle \hat{V}^1_{\iota,j}(t)\left[\hat{O}^{(s)}(t), \hat{D}^-_m(t)\right]\hat{V}^{\iota}_{2,l}(t)\right\rangle \\
&+ \frac{i}{2\tau^s_{sad}} \sum_{m,j,l=1}\left\{U_s(j,m,l)\langle[\hat{O}^{(s)}(t), \hat{V}^1_{\iota,j}(t)]\hat{V}^{\iota}_{2l}(t)\hat{D}^-_m(t)\rangle \right. \\
&\left. + U^*_s(j,m,l)\langle \hat{V}^1_{\iota,j}(t)\hat{D}^-_m(t)[\hat{O}^{(s)}(t), \hat{V}^{\iota}_{2,l}(t)]\rangle\right\} + H.c.
\end{aligned}
\tag{23}
$$

The similar expression can be obtained for a Λ three-level system in interaction with D radiators, doing the substitution $\hat{V}^{\beta}_{\alpha,j} \to \Lambda^{\alpha}_{\beta j}$. For two atoms in each subsystem, an attractive peculiarity follows from this substitution. If $O_s(t)$ is the inversion of the D atom, the direct modification of the D atomic excitation by Λ three-level atom is equal to zero $\left\langle \hat{\Lambda}^{\iota}_{1,l}(t)\hat{\Lambda}^2_{\iota,l}(t)\left[\hat{N}_d(t), \hat{D}^-_m(t)\right]\right\rangle = 0$ due to the operator product $\hat{\Lambda}^{\iota}_{1,l}(t)\hat{\Lambda}^2_{\iota,l}(t) = 0$ for the same atom. In order to obtain the closed system of equation from master Eqs. (15)and (23), we consider the simple interaction of two atoms in the scattering process represented by the analytical scheme of **Figure 1B**. In this case we introduce the new indexes "s" and "a" instead of "1" and "2", which correspond to the Stokes and anti-Stokes scattering frequencies ω_s and ω_a. Considering that the anti-Stokes frequency ω_a is larger than Stokes ω_s, one can approximate the exchange integrals (17) with expression

$$
V_s \simeq \frac{\sin(x_a)}{x_a} + i\frac{1 - \cos(x_a)}{x_a}. \tag{24}
$$

Here $x_a = \omega_a r/c$. The mean values of the operators $\langle \hat{N}_s\rangle = \left\langle \hat{V}^2_2\right\rangle$, $\langle \hat{N}_a\rangle = \left\langle \hat{V}^1_1\right\rangle$, and $\langle \hat{N}_d\rangle$ are considered the populations of excited states of V and D radiators,

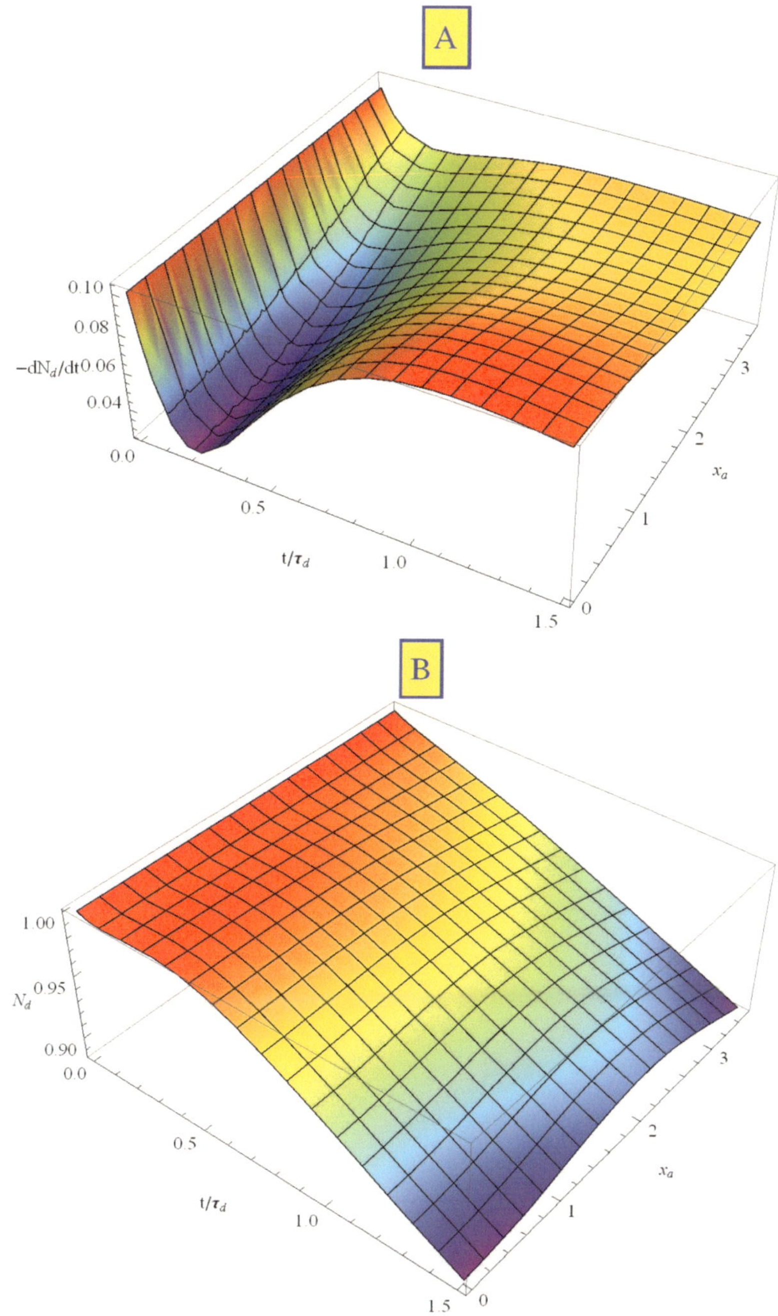

Figure 5.
The decay process of the dipole-forbidden transitions of the D radiator under the influence of V three-level radiator for following parameter atom for following parameters of the system, $\langle N_a \rangle = 0.5$, $\langle N_s \rangle = 0.5$, $\langle N_d \rangle = 1$, $\tau_a/\tau_d = 0.1$, $\tau_a/\tau_s = 6$, *and* τ_a/τ_{asd}, *(A) represents the decay rate* $-d\langle N_d/dt \rangle$ *and (B) represents the excitation of the D atom plotted as the numerical solution of the system of Eq. (25) as function of* t/τ_a*) and relative distance* $x_a = 2\pi r/\lambda_a$ *in the three-dimensional representation.*

respectively. The functions $\langle \hat{F}_s(t, x_a) \rangle = i\left[\left\langle \hat{V}_2^1(t)\hat{D}^-(t)\right\rangle - \left\langle \hat{D}^+\hat{V}_1^2(t)\right\rangle\right]$, $\langle \hat{E}_s(t, x_a) \rangle = \left[\left\langle \hat{V}_2^1(t)\hat{D}^-(t)\right\rangle - \left\langle \hat{D}^+(t)\hat{V}_1^2(t)\right\rangle\right]$, $\langle \hat{N}_d \hat{N}_s \rangle$, and $\langle \hat{N}_d \hat{N}_a \rangle$ describe the polarization and population correlations between the atoms Ξ and D. For this two-atom system, we can obtain the following closed system of equations from generalized equation (23).

$$
\begin{aligned}
\frac{d}{dt}\langle\hat{N}_\iota(t,x_a)\rangle &= \frac{1}{\tau_{\iota,a}}\langle\hat{N}_a(t,x_a)\rangle + \frac{1}{\tau_{\iota,s}}\langle\hat{N}_s(t,x_a)\rangle + \frac{1}{\tau^s_{sad}}\frac{1-\cos(x_a)}{x_a}\langle\hat{E}_s(t,x_a)\rangle,\\
\frac{d}{dt}\langle\hat{N}_a(t,x_a)\rangle &= -\frac{1}{\tau_{\iota,a}}\langle\hat{N}_a(t,x_a)\rangle\\
&\quad+\frac{1}{2\tau^s_{sad}}\left[\frac{\sin(x_a)}{x_a}\langle\hat{F}_s(t,x_a)\rangle - \frac{1-\cos(x_a)}{x_a}\langle\hat{E}_s(t,x_a)\rangle\right],\\
\frac{d}{dt}\langle\hat{N}_s(t,x_a)\rangle &= -\frac{1}{\tau_{\iota,s}}\langle\hat{N}_s(t,x_a)\rangle\\
&\quad-\frac{1}{2\tau^s_{sad}}\left[\frac{\sin(x_a)}{x_a}\langle\hat{F}_s(t,x_a)\rangle + \frac{1-\cos(x_a)}{x_a}\langle\hat{E}_s(t,x_a)\rangle\right],\\
\frac{d}{dt}\langle\hat{N}_d(t,x_a)\rangle &= -\frac{1}{\tau_d}\langle\hat{N}_d(t,x_a)\rangle\\
&\quad+\frac{1}{2\tau^s_{sad}}\left[\frac{\sin(x_a)}{x_a}\langle\hat{F}_s(t,x_a)\rangle - \frac{1-\cos(x_a)}{x_a}\langle\hat{E}_s(t,x_a)\rangle\right],\\
\frac{d}{dt}\langle\hat{F}_s(t,x_a)\rangle &= -\frac{1}{2}\left\{\frac{1}{\tau_d}+\frac{1}{\tau_{\iota,s}}+\frac{1}{\tau_{\iota,a}}\right\}\langle\hat{F}_s(t,x_a)\rangle\\
&\quad+\frac{1}{\tau^s_{sad}}\frac{\sin(x_a)}{x_a}\left[\langle\hat{N}_d(t,x_a)\hat{N}_s(t,x_a)\rangle - \langle(1-\hat{N}_d(t,x_a))\hat{N}_a(t,x_a)\rangle\right],\\
\frac{d}{dt}\langle\hat{E}_s(t,x_a)\rangle &= -\frac{1}{2}\left\{\frac{1}{\tau_d}+\frac{1}{\tau_{\iota,s}}+\frac{1}{\tau_{\iota,a}}\right\}\langle\hat{E}_s(t,x_a)\rangle\\
&\quad-\frac{1-\cos(x_a)}{x_a\tau^s_{sad}}\left\{\langle\hat{N}_d(t,x_a)\hat{N}_s(t,x_a)\rangle + \langle(1-\hat{N}_d(t,x_a))\hat{N}_a(t,x_a)\rangle\right\}\\
\frac{d}{dt}\langle\hat{N}_d(t,x_a)\hat{N}_s(t,x_a)\rangle &= -\left[\frac{1}{\tau_d}+\frac{1}{\tau_{\iota,s}}\right]\langle\hat{N}_d(t,x_a)\hat{N}_s(t,x_a)\rangle\\
&\quad-\frac{1}{2\tau^s_{sad}}\left[\frac{\sin(x_a)}{x_a}\langle\hat{F}_s(t,x_a)\rangle + \frac{1-\cos(x_a)}{x_a}\langle\hat{E}_s(t,x_a)\rangle\right],\\
\frac{d}{dt}\langle\hat{N}_d(t,x_a)\hat{N}_a(t,x_a)\rangle &= -\left[\frac{1}{\tau_d}+\frac{1}{\tau_{\iota a}}\right]\langle\hat{N}_d(t,x_a)\hat{N}_a(t,x_a)\rangle.
\end{aligned}
\tag{25}
$$

As follows from the system (25), and numerical simulation plotted in **Figure 5** the first $\langle\hat{N}_d\rangle/\tau_d$ and second terms $(1/\tau^s_{sad})\langle\hat{F}_s\rangle$ describe the generation rate of entangled photon pairs and scattering rate with absorption of Stokes photon and generation of two anti-Stokes photons by the system formed from V and D atoms. When the time tends to infinity, all excited atomic energies $E_0 = \hbar\omega_a + \hbar\omega_s + \hbar\omega_d$ of three-level V and two-level D atoms are emitted by the system. Taking into account the conservation law in the scattering process $\omega_a - \omega_s - \omega_d = 0$, we observe that this cooperation between the atoms becomes predominant, when the collective scattering rate $1/\tau^s_{sad}$ increases. In other words, the probability of absorption of Stokes photon $\hbar\omega_s$ which is accompanied with the generation of the new anti-Stokes photon $\hbar\omega_a$ by D atom becomes possible. In this case two atoms represented in the **Figure 1B** can generate an entangled anti-Stokes photons with energy $E_0 = 2\hbar\omega_a$. The possibility of the excitation transfer between the atoms Ξ and D represented in **Figure 4B** can be found in the special preparation of the system.

We can conclude that it is possible to study all cooperations two-photon process between single atoms in each system represented in **Figure 1A–C**. For example, the system of Eqs. (20) and (25) can be solved simultaneously taking into consideration scattering and two-photon transitions. In this case the effective energy transfer of the excitation between the atoms Ξ, V, and D radiator prepared in the special initial states can open the new possibilities of non-resonance interaction between the atomic subsystems.

4. Conclusions

This chapter proposed the cooperative effects between three-level system and dipole-forbidden two-level systems in nonlinear interaction through the vacuum field during the spontaneous emission time. The possibility of cooperative migration of energy from one excited dipole-active three-level atom to another takes place with phase retardation effects and depends on the position of atoms in the system. This excitation transfer from dipole-active to dipole-forbidden subsystems takes place with phase dependence amplitudes, so that the cooperative excitation of the system consisted from two species of atoms depends on the retardation of radiation along the sample and geometry of the system. This follows from the excited or ground state of one of the radiators represented in **Figures 4** and **5**. As in Ref. [20], the exchanges between the Ξ (or V) three-level atom and D take place with the absorption and emission of two quanta, but in this chapter, we take into consideration the real and imaginary parts of exchange integrals. In this case, two correlation functions introduced functions $\langle \hat{F}_{b(s)}(t, x_a) \rangle$ and $\langle \hat{E}_{b(s)}(t, x_a) \rangle$, which modify the dynamics of possible excitation of D atoms by Ξ and V radiators. The scattering transfer of the energy between the excited state of V three-level radiator and dipole-forbidden transitions of D two-level atoms are effective when the dipole-forbidden atom enters in the two-photon resonance with the energy difference between the two dipole transitions (**Figures 1A** and **5A**). When the atom D is in the excited state, the emitted Stokes photon by one atom of the V systems can be absorbed by another radiator from the D subsystem, so that two radiators pass into the ground state generating two anti-Stokes photons with energies $E_0 = 2\hbar\omega_a$. The opposite situation can be observed when D atom is prepared in the ground state.

Author details

Nicolae A. Enaki
Quantum Optics and Kinetic Processes Lab, Institute of Applied Physics of Moldova, Chisinau, Republic of Moldova

*Address all correspondence to: enakinicolae@yahoo.com

References

[1] Dicke RH. Coherence in spontaneous radiation processes. Physics Review. 1954;**93**:99-110

[2] Florian R, Schwan LO, Schmid D. Superradiance and high-gain mirrorless laser activity of O_2^--centers in KCl. Solid State Communications. 1982;**42**: 55-57

[3] Rautian SG, Chernobrod BM. Cooperative effect in Raman scattering of light. Soviet Physics—JETP. 1977;**45**: 705-708

[4] Inouye S, Pfau T, Gupta S, Chikkatur AP, Görlitz A, Pritchard DE, et al. Phase-coherent amplification of atomic matter waves. Nature. 1999;**402**: 641-644

[5] Enaki NA. Superradiation from two-photon spontaneous decay. Soviet Physics—JETP. 1988;**67**:2033-2038

[6] Yuan L, Hokr BH, Traverso AJ, et al. Theoretical analysis of the coherence-brightened laser in air. Physical Review A. 2013;**023826**:87

[7] Schwendimann P. Damping effects in two-colour superfluorescence. Optica Acta: International Journal of Optics. 1984;**31**:107-114

[8] Haake F, Reibold R. Two-color superfluorescence from three-level systems. Physics Letters A. 1982;**92**: 29-31

[9] Andreev AV, Enaki NA, Ilinskii YA. Superfluorescence in a three-level system. Theoretical and Mathematical Physics. 1985;**64**:960-965

[10] Pando Lambruschini CL. A semiclassical approach for two-color four-mode solid-state superfluorescence. Optics Communications. 1991;**85**: 291-298

[11] Traverso AJ, Sanchez-Gonzalez R, Yuan L, et al. Coherence brightened laser source for atmospheric remote sensing. Proceedings of the National Academy of Sciences of the United States of America. 2012;**109**:15185-15190

[12] Enaki NA. Mutual cooperative effects between single- and two-photon super-fluorescent processes through vacuum field. European Physical Journal D: Atomic, Molecular, Optical and Plasma Physics. 2012;**66**(98)

[13] Enaki NA. Non-Linear Cooperative Effects in Open Quantum Systems: Entanglement and Second Order Coherence. NY, USA: Nova Science Publishers; 2015. p. 325

[14] Bonifacio R, Lugiato LA. Cooperative radiation processes in two-level systems: Superfluorescence. Physical Review A. 1974;**11**:1507-1521

[15] Lehmberg RH. Radiation from an N-atom system. I. General formalism. Physical Review A. 1970;**2**:883

[16] Lugiato LA, Oldano C, Narducci CLM. Cooperative frequency locking and stationary spatial structures in lasers. Journal of the Optical Society of America B: Optical Physics. 1988;**5**:879

[17] Wang T, Yelin SF, Eyler1 RCEE, Farooqi SM, Gould PL, Kostrun M, et al. Superradiance in ultracold Rydberg gases. Physical Review A. 2007;**75** (033802)

[18] Paradis E, Barrett B, Kumarakrishnan A, Zhang R, Raithel G. Observation of superfluorescent emissions from laser-cooled atoms. Physical Review A. 2008;**77**:043419

[19] Enaki NA. The photon quantum statistics in the processes of two quanta optical nutation. Soviet Physics.

Žurnal ėksperimental'noj i teoretičeskoj fiziki. 1990;**98**:783-796

[20] Enaki NA, Rosca T. Cooperative effects between three subsystems in two-photon and Raman resonances. In: Proceedings of the SPIE 8882, ROMOPTO 2012: Tenth conference on optics: Micro- to Nanophotonics III, 88820L; 2013. 11 p

Chapter 3

Processes of Creation and Propagation of Correlations in Large Quantum Particle System

Viktor I. Gerasimenko

Abstract

We review new approaches to the description of the evolution of states of large quantum particle systems by means of the marginal correlation operators. Using the definition of marginal correlation operators within the framework of dynamics of correlations governed by the von Neumann hierarchy, we establish that a sequence of such operators is governed by the nonlinear quantum BBGKY hierarchy. The constructed nonperturbative solution of the Cauchy problem to this hierarchy of nonlinear evolution equations describes the processes of the creation and the propagation of correlations in large quantum particle systems. Furthermore, we consider the problem of the rigorous description of collective behavior of quantum many-particle systems by means of a one-particle (marginal) correlation operator that is a solution of the generalized quantum kinetic equation with initial correlations, in particular, correlations characterizing the condensed states of systems.

Keywords: von Neumann hierarchy, nonlinear quantum BBGKY hierarchy, quantum kinetic equation, correlation of states, scaling limit
2000 Mathematics Subject Classification: 35Q40; 47D06

1. Introduction

In this chapter, we consider mathematical problems concerning the description of processes of a creation and a propagation of correlations in quantum many-particle systems, namely, correlations in quantum systems both finitely and infinitely many particles and the description of correlations by means of the state of typical particle of large quantum particle system.

As known, the marginal correlation operators give an equivalent approach to the description of the evolution of states of quantum systems of many particles in comparison with marginal density operators [1]. The physical interpretation of marginal correlation operators is that the macroscopic characteristics of fluctuations of mean values of observables are determined by them on the microscopic level [1, 2].

Traditionally marginal correlation operators are introduced by means of the cluster expansions of the marginal density operators [2–4]. In articles [5, 6] an approach based on the definition of the marginal correlation operators within the framework of dynamics of correlations governed by the von Neumann hierarchy was developed. As a result of which, it is established that the marginal correlation operators are governed by the hierarchy of nonlinear evolution equations, known as the quantum nonlinear BBGKY (Bogoliubov-Born-Green-Kirkwood-Yvon) hierarchy, and its solution is represented in the form of series, the generating operator of every term of which are the corresponding-order cumulant of groups of nonlinear operators of the von Neumann hierarchy for correlation operators [7].

In the chapter, we also consider the problem of the rigorous description of the evolution of correlations in quantum many-particle systems by means of a one-particle (marginal) density operator that is a solution of the generalized quantum kinetic equation with initial correlations [8]. We remark that initial states specified by correlations are typical for the condensed states of many-particle systems in contrast to their gaseous state [1].

We note that in modern researches, the conventional approach to the problem of the rigorous derivation of kinetic equations lies in the construction of various scaling limits of a solution of equations, describing the evolution of the state of many-particle systems [9], in particular, a mean field limit of a perturbative solution of the BBGKY hierarchy for a sequence of marginal density operators [10–17].

2. Dynamics of quantum correlations

As known [1, 2], quantum systems of fixed number of particles are described in terms of observables and states. The functional of the mean value of observables defines a duality between observables and states, and as a consequence, there exist two approaches to the description of the evolution of quantum systems, namely, in terms of observables that are governed by the Heisenberg equation and in terms of states governed by the von Neumann equation for the density operator, respectively. An equivalent approach to the description of states of quantum systems is given by means of operators determined by the cluster expansions of the density operator which are interpreted as correlation operators. In this section we consider fundamental equations describing the evolution of correlations of quantum systems with a finite number of particles.

2.1 Preliminaries

We denote by $\mathcal{F}_{\mathcal{H}} = \oplus_{n=0}^{\infty}\mathcal{H}^{\otimes n}$ the Fock space over the Hilbert space $\mathcal{H}$, where $\mathcal{H}^{\otimes n} \equiv \mathcal{H}_n$ is the n-particle Hilbert space. Let $\mathfrak{L}^1(\mathcal{H}_n)$ be the space of trace class operators $f_n \equiv f_n(1, \ldots, n) \in \mathfrak{L}^1(\mathcal{H}_n)$ that satisfy the symmetry condition $f_n(1, \ldots, n) = f_n(i_1, \ldots, i_n)$ for arbitrary $(i_1, \ldots, i_n) \in (1, \ldots, n)$ and are equipped with the norm

$$\|f_n\|_{\mathfrak{L}^1(\mathcal{H}_n)} = \mathrm{Tr}_{1,\ldots,n}|f_n(1, \ldots, n)|,$$

where $\mathrm{Tr}_{1,\ldots,n}$ are partial traces over $1, \ldots, n$ particles. We denote by $\mathfrak{L}^1_0(\mathcal{H}_n)$ the everywhere dense set of finite sequences of degenerate operators with infinitely differentiable kernels with compact supports.

On the space of trace class operators $\mathfrak{L}^1(\mathcal{H}_n)$, it is defined as the one-parameter mapping $\mathcal{G}^*_n(t)$

$$\mathbb{R}^1 \ni t \mapsto \mathcal{G}^*_n(t)f_n \doteq e^{-itH_n}f_n e^{itH_n}, \tag{1}$$

where the following units are used: $m = 1$ is the mass of a particle, $h = 2\pi\hbar = 1$ is a Planck constant, and the self-adjoint operator H_n is the Hamiltonian of n particles, obeying Maxwell-Boltzmann statistics. Further an inverse group to group (1) will be denoted by $(\mathcal{G}_n^*)^{-1}(t) = \mathcal{G}_n^*(-t)$.

On its domain of the definition, the infinitesimal generator $\mathcal{N}_n^*$ of the group of operators (1) is determined in the sense of the strong convergence of the space $\mathfrak{L}^1(\mathcal{H}_n)$ by the operator

$$\lim_{t\to 0}\frac{1}{t}\left(\mathcal{G}_n^*(t)f_n - f_n\right) = -i\left(H_n f_n - f_n H_n\right) \doteq \mathcal{N}_n^* f_n, \tag{2}$$

that has the following structure: $\mathcal{N}_n^* = \sum_{j=1}^n \mathcal{N}^*(j) + \epsilon\sum_{j_1<j_2=1}^n \mathcal{N}_{\text{int}}^*(j_1,j_2)$, where the operator $\mathcal{N}^*(j)$ is a free motion generator of the von Neumann equation (the dual operator to the generator of the Heisenberg equation for observables) [2], the operator $\mathcal{N}_{\text{int}}^*$ is defined by means of the operator of a two-body interaction potential Φ by the formula $\mathcal{N}_{\text{int}}^*(j_1,j_2)f_n \doteq -i\left(\Phi(j_1,j_2)f_n - f_n\Phi(j_1,j_2)\right)$, and we denoted a scaling parameter by $\epsilon > 0$.

Let the symbol $\sum_{\mathrm{P}:(1,\dots,s)=\cup_j X_j}$ denote the sum over all possible partitions P of the set $(1,\dots,s)$ into $|\mathrm{P}|$ nonempty mutually disjoint subsets X_j, and the set $(\{X_1\},\dots,\{X_{|\mathrm{P}|}\})$ consists from elements which are subsets $X_j \subset (1,\dots,s)$ of the set $(1,\dots,s)$, i.e., $|(\{X_1\},\dots,\{X_{|\mathrm{P}|}\})| = |\mathrm{P}|$. On the space $\mathfrak{L}^1(\mathcal{F}_\mathcal{H}) = \oplus_{n=0}^\infty \mathfrak{L}^1(\mathcal{H}_n)$ of sequences $f = (f_0, f_1, \dots, f_n, \dots)$ of trace class operators $f_n \in \mathfrak{L}^1(\mathcal{H}_n)$ and $f_0 \in \mathbb{C}$, the following nonlinear one-parameter mapping is defined:

$$\mathcal{G}(t;1,\dots,s|f) \doteq \sum_{\mathrm{P}:(1,\dots,s)=\cup_j X_j} \mathfrak{A}_{|\mathrm{P}|}\left(t,\{X_1\},\dots,\{X_{|\mathrm{P}|}\}\right)\prod_{X_j\subset \mathrm{P}} f_{|X_j|}(X_j), \quad s\geq 1, \tag{3}$$

where the generating operator $\mathfrak{A}_{|\mathrm{P}|}(t)$ of this expansion is the $|\mathrm{P}|th$-order cumulant of the groups of operators (1) defined by the following expansion [2]:

$$\mathfrak{A}_{|\mathrm{P}|}\left(t,\{X_1\},\dots,\{X_{|\mathrm{P}|}\}\right) \doteq \sum_{\mathrm{P}':(\{X_1\},\dots,\{X_{|\mathrm{P}|}\})=\cup_k Z_k} (-1)^{|\mathrm{P}'|-1}(|\mathrm{P}'|-1)!\prod_{Z_k\subset \mathrm{P}'} \mathcal{G}_{|\theta(Z_k)|}^*(t,\theta(Z_k)), \tag{4}$$

and θ is the declusterization mapping: $\theta\left(\{X_1\},\dots,\{X_{|\mathrm{P}|}\}\right) \doteq (1,\dots,s)$.

Below we adduce the examples of mapping expansions (3):

$$\begin{aligned}
&\mathcal{G}(t;1|f) = \mathfrak{A}_1(t,1)f_1(1),\\
&\mathcal{G}(t;1,2|f) = \mathfrak{A}_1(t,\{1,2\})f_2(1,2) + \mathfrak{A}_{1+1}(t,1,2)f_1(1)f_1(2),\\
&\mathcal{G}(t;1,2,3|f) = \mathfrak{A}_1(t,\{1,2,3\})f_3(1,2,3) + \mathfrak{A}_{1+1}(t,1,\{2,3\})f_1(1)f_2(2,3)+\\
&\mathfrak{A}_{1+1}(t,2,\{1,3\})f_1(2)f_2(1,3) + \mathfrak{A}_{1+1}(t,3,\{1,2\})f_1(3)f_2(1,2)+\\
&\mathfrak{A}_3(t,1,2,3)f_1(1)f_1(2)f_1(3).
\end{aligned}$$

For $f_s \in \mathfrak{L}^1(\mathcal{H}_s)$, $s\geq 1$, the mapping $\mathcal{G}(t;1,\dots,s|f)$ is defined, and, according to the inequality

$$\|\mathfrak{A}_{|\mathrm{P}|}\left(t,\{X_1\},\dots,\{X_{|\mathrm{P}|}\}\right)f_s\|_{\mathfrak{L}^1(\mathcal{H}_s)} \leq |\mathrm{P}|!\, e^{|\mathrm{P}|}\|f_s\|_{\mathfrak{L}^1(\mathcal{H}_s)},$$

the following estimate is true:

$$\|\mathcal{G}(t;1,\dots,s|f)\|_{\mathfrak{L}^1(\mathcal{H}_s)} \leq s!\, e^{2s}c^s, \tag{5}$$

where $c \equiv e^3\max\left(1, \max_{\mathrm{P}:(1,\dots,s)=\cup_i X_i}\|f_{|X_i|}\|_{\mathfrak{L}^1(\mathcal{H}_{|X_i|})}\right)$. On the space $\mathfrak{L}^1(\mathcal{F}_\mathcal{H})$, one-parameter mapping (3) is a bounded strong continuous group of nonlinear operators.

2.2 The von Neumann hierarchy for correlation operators

The evolution of all possible states of a quantum system of non-fixed, i.e., arbitrary but finite, number of identical particles, obeying the Maxwell-Boltzmann statistics, can be described by means of the sequence $g(t) = (g_0, g_1(t), \ldots, g_s(t), \ldots) \in \mathfrak{L}^1(\mathcal{F}_\mathcal{H})$ of the correlation operators $g_s(t) = g_s(t, 1, \ldots, s)$, $s \geq 1$, governed by the Cauchy problem of the von Neumann hierarchy [5]:

$$\frac{\partial}{\partial t} g_s(t, 1, \ldots, s) = \mathcal{N}_s^* g_s(t, 1, \ldots, s) + \epsilon \sum_{\mathrm{P}:(1,\ldots,s) = X_1 \cup X_2} \sum_{i_1 \in X_1} \sum_{i_2 \in X_2} \mathcal{N}_{\mathrm{int}}^*(i_1, i_2) g_{|X_1|}(t, X_1) g_{|X_2|}(t, X_2), \tag{6}$$

$$g_s(t)\big|_{t=0} = g_s^{0,\epsilon}, \quad s \geq 1, \tag{7}$$

where $\epsilon < 0$ is a scaling parameter, the symbol $\sum_{\mathrm{P}:(1,\ldots,s) = X_1 \cup X_2}$ means the sum over all possible partitions P of the set $(1, \ldots, s)$ into two nonempty mutually disjoint subsets X_1 and X_2, and the operator $\mathcal{N}_s^*$ is defined on the subspace $\mathfrak{L}_0^1(\mathcal{H}_s)$ by formula (2).

We remark that correlation operators can be introduced by means of the cluster expansions [2] of the density operators (the kernel of a density operator is known as a density matrix) governed by a sequence of the von Neumann equations, and hence, they describe the evolution of states by an equivalent method in comparison with the density operators. For quantum systems of fixed number of particles, the state is described by finite sequence of correlation operators governed by a corresponding system of the von Neumann equations (6).

A solution (nonperturbative solution) of the Cauchy problem of the von Neumann hierarchy for correlation operators (6) and (7) is represented by group of nonlinear operators (3)

$$g(t, 1, \ldots, s) = \mathcal{G}(t; 1, \ldots, s | g(0)), \quad s \geq 1, \tag{8}$$

where a sequence of initial correlation operators (7) is denoted by $g(0) = (g_0, g_1^{0,\epsilon}, \ldots, g_n^{0,\epsilon}, \ldots)$ and $g_0 \in \mathbb{C}$.

We remark, if at initial time there are no correlations between particles, i.e., in the case of initial states, satisfying a chaos condition [2], a sequence of initial correlation operators takes the form $g(0) = (0, g_1^{0,\epsilon}, 0, \ldots, 0, \ldots)$. Then solution (8) of the Cauchy problem of the von Neumann hierarchy (6) and (7) is represented by the following expansions:

$$g_s(t, 1, \ldots, s) = \mathfrak{A}_s(t, 1, \ldots, s) \prod_{i=1}^{s} g_1^{0,\epsilon}(i), \quad s \geq 1,$$

where the operator $\mathfrak{A}_s(t)$ is the sth-order cumulant of groups of operators (1) determined by the expansion

$$\mathfrak{A}_s(t, 1, \ldots, s) = \sum_{\mathrm{P}:(1,\ldots,s) = \cup_i X_i} (-1)^{|\mathrm{P}|-1}(|\mathrm{P}| - 1)! \prod_{X_i \subset \mathrm{P}} \mathcal{G}_{|X_i|}^*(t, X_i), \tag{9}$$

and we used notations accepted in formula (3).

We remark also that nonperturbative solution (8) of the Cauchy problem of the von Neumann hierarchy (6) and (7) can be transformed to the perturbation (iteration) expansion as a result of the application of analogs of the Duhamel equation to cumulants (4) of groups of operators (1).

The following statement is true [6]. In the case of bounded interaction potentials for $t \in \mathbb{R}$, a solution of the Cauchy problem of the von Neumann hierarchy (6) and (7) is determined by a sequence of correlation operators represented by formula (8). If $g_n^{0,\epsilon} \in \mathfrak{L}_0^1(\mathcal{H}_n) \subset \mathfrak{L}^1(\mathcal{H}_n)$, it is a strong solution, and for arbitrary initial data $g_n^{0,\epsilon} \in \mathfrak{L}^1(\mathcal{H}_n)$, it is a weak solution.

The stated above results can be extended to quantum systems of bosons and fermions like in paper [6].

3. The evolution of correlations in large quantum particle systems

An equivalent approach in describing the states of quantum systems of many particles consists in describing states by means of marginal density operators governed by the BBGKI hierarchy or by means of operators determined by their cluster expansions, which are interpreted as marginal correlation operators [1]. On the microscopic scale, the macroscopic characteristics of fluctuations of observables are directly determined by the marginal correlation operators. Such approach allows us to describe the evolution of correlations in quantum systems both with finite and infinite number of particles.

3.1 The hierarchy of evolution equations for marginal correlation operators

Traditionally marginal correlation operators are determined by means of the cluster expansions of the marginal density operators [2–4]. We introduce the marginal correlation operators in the framework of the solution of the Cauchy problem for the von Neumann hierarchy (6) and (7) by the following series expansions:

$$G_s(t,1,\dots,s) \doteq \sum_{n=0}^{\infty} \frac{1}{n!} \mathrm{Tr}_{s+1,\dots,s+n}\, \mathcal{G}(t;1,\dots,s+n|g(0)), \quad s \geq 1. \tag{10}$$

According to estimate (5), series (10) exists and the following estimate holds: $\|G_s(t)\|_{\mathfrak{L}^1(\mathcal{H}_s)} \leq s!(2e^2)^s c^s \sum_{n=0}^{\infty}(2e^2)^n c^n$, where $c \equiv e^3 \max\left(1, \max_{\mathrm{P}:(1,\dots,s)=\cup_i X_i} \|g_{|X_i|}(0)\|_{\mathfrak{L}^1(\mathcal{H}_{|X_i|})}\right)$.

We remark that the macroscopic characteristics of fluctuations of observables are directly determined by marginal correlation operators (10), for example, the functional of the dispersion of the additive-type observables, i.e., $A^{(1)} = \left(0, a_1(1), \dots, \sum_{i_1=1}^{n} a_1(i_1), \dots\right)$, is represented by the formula [1]

$$\begin{aligned}\left\langle \left(A^{(1)} - \left\langle A^{(1)} \right\rangle\right)^2 \right\rangle(t) = &\mathrm{Tr}_1 \left(a_1^2(1) - \left\langle A^{(1)} \right\rangle^2(t)\right) G_1(t,1) \\ &+ \mathrm{Tr}_{1,2}\, a_1(1) a_1(2) G_2(t,1,2),\end{aligned}$$

where $\left\langle A^{(1)} \right\rangle(t) = \mathrm{Tr}_1 a_1(1) G_1(t,1)$ is a mean-value functional of the additive-type observable [2].

Then the evolution of all possible states of large quantum particle systems, obeying the Maxwell-Boltzmann statistics, can be described by means of the sequence $G(t) = (I, G_1(t), G_2(t), \dots, G_s(t), \dots) \in \mathfrak{L}^1(\mathcal{F}_\mathcal{H})$ of marginal correlation operators governed by the Cauchy problem of the following hierarchy of nonlinear evolution equations (the nonlinear quantum BBGKY hierarchy):

$$\frac{\partial}{\partial t}G_s(t,1,\ldots,s) = \mathcal{N}^*_s G_s(t,1,\ldots,s) +$$
$$\epsilon \sum_{\mathrm{P}:(1,\ldots,s)=X_1\cup X_2} \sum_{i_1\in X_1}\sum_{i_2\in X_2} \mathcal{N}^*_{\mathrm{int}}(i_1,i_2)G_{|X_1|}(t,X_1)G_{|X_2|}(t,X_2)) +$$
$$\epsilon \mathrm{Tr}_{s+1}\sum_{i\in Y}\mathcal{N}^*_{\mathrm{int}}(i,s+1)(G_{s+1}(t,1,\ldots,s+1) + \tag{11}$$
$$\sum_{\substack{\mathrm{P}:\,(1,\ldots,s+1)=X_1\cup X_2,\\ i\in X_1;\, s+1\in X_2}} G_{|X_1|}(t,X_1)G_{|X_2|}(t,X_2)),$$

$$G_s(t)\big|_{t=0} = G_s^{0,\epsilon}, \quad s\geq 1, \tag{12}$$

where $\epsilon > 0$ is a scaling parameter and we use accepted in hierarchy (6) notations.

If $G(0) = \left(I, G_1^{0,\epsilon}(1),\ldots,G_s^{0,\epsilon}(1,\ldots,s),\ldots\right)$ is a sequence of initial marginal correlation operators (12), then a nonperturbative solution of the Cauchy problem (11) and (12) is represented by the following sequence of self-adjoint operators:

$$G_s(t,1,\ldots,s) = \sum_{n=0}^{\infty}\frac{1}{n!}\mathrm{Tr}_{s+1,\ldots,s+n}\mathfrak{A}_{1+n}(t;\{1,\ldots,s\},s+1,\ldots,s+n|G(0)), \quad s\geq 1, \tag{13}$$

where the generating operator $\mathfrak{A}_{1+n}(t;\{1,\ldots,s\},s+1,\ldots,s+n|G(0))$ of this series is the $(1+n)th$-order cumulant of groups of nonlinear operators (3):

$$\mathfrak{A}_{1+n}(t;\{1,\ldots,s\},s+1,\ldots,s+n|G(0)) \doteq$$
$$\sum_{\mathrm{P}:(\{1,\ldots,s\},s+1,\ldots,s+n)=\cup_k X_k} (-1)^{|\mathrm{P}|-1}(|\mathrm{P}|-1)!\mathcal{G}(t;\theta(X_1)|\ldots\mathcal{G}(t;\theta(X_{|\mathrm{P}|})|G(0))\ldots), \quad n\geq 0, \tag{14}$$

and composition of mappings (3) of the corresponding noninteracting groups of particles we denote by $\mathcal{G}(t;\theta(X_1)|\ldots\mathcal{G}(t;\theta(X_{|\mathrm{P}|})|G(0))\ldots)$, for example,

$$\mathcal{G}(t;1|\mathcal{G}(t;2|f)) = \mathfrak{A}_1(t,1)\mathfrak{A}_1(t,2)f_2(1,2),$$
$$\mathcal{G}(t;1,2|\mathcal{G}(t;3|f)) = \mathfrak{A}_1(t,\{1,2\})\mathfrak{A}_1(t,3)f_3(1,2,3) +$$
$$\mathfrak{A}_2(t,1,2)\mathfrak{A}_1(t,3)\left(f_1(1)f_2(2,3)+f_1(2)f_2(1,3)\right).$$

Below we adduce the examples of expansions (14). The first-order cumulant of the groups of nonlinear operators (3) is the same group of nonlinear operators, i.e.,

$$\mathfrak{A}_1(t;\{1,\ldots,s\}|G(0)) = \mathcal{G}(t;1,\ldots,s|G(0)).$$

In the case of $s=2$, the second-order cumulant of nonlinear operators (3) has the structure

$$\mathfrak{A}_{1+1}(t;\{1,2\},3|G(0)) = \mathcal{G}(t;1,2,3|G(0)) - \mathcal{G}(t;1,2|\mathcal{G}(t;3|G(0))) =$$
$$\mathfrak{A}_{1+1}(t,\{1,2\},3)G_3^{0,\epsilon}(1,2,3) + (\mathfrak{A}_{1+1}(t,\{1,2\},3) - \mathfrak{A}_{1+1}(t,2,3)\mathfrak{A}_1(t,1))G_1^{0,\epsilon}(1)G_2^{0,\epsilon}(2,3) +$$
$$(\mathfrak{A}_{1+1}(t,\{1,2\},3) - \mathfrak{A}_{1+1}(t,1,3)\mathfrak{A}_1(t,2))G_1^{0,\epsilon}(2)G_2^{0,\epsilon}(1,3) +$$
$$\mathfrak{A}_{1+1}(t,\{1,2\},3)G_1^{0,\epsilon}(3)G_2^{0,\epsilon}(1,2) + \mathfrak{A}_3(t,1,2,3)G_1^{0,\epsilon}(1)G_1^{0,\epsilon}(2)G_1^{0,\epsilon}(3),$$

where the operator

$$\mathfrak{A}_3(t,1,2,3) = \mathfrak{A}_{1+1}(t,\{1,2\},3) - \mathfrak{A}_{1+1}(t,2,3)\mathfrak{A}_1(t,1) - \mathfrak{A}_{1+1}(t,1,3)\mathfrak{A}_1(t,2)$$

is the third-order cumulant (9) of groups of operators (1).

In the case of initial data specified by the sequence of marginal correlation operators

$$G^{(c)} = \left(0, G_1^{0,\epsilon}, 0, \dots, 0, \dots\right), \tag{15}$$

i.e., initial states satisfying a chaos property [9], according to definition (14), marginal correlation operators (13) are represented by the following series expansions:

$$G_s(t, 1, \dots, s) = \sum_{n=0}^{\infty} \frac{1}{n!} \mathrm{Tr}_{s+1,\dots,s+n} \mathfrak{A}_{s+n}(t; 1, \dots, s+n) \prod_{i=1}^{s+n} G_1^{0,\epsilon}(i), \quad s \geq 1, \tag{16}$$

where the generating operator $\mathfrak{A}_{s+n}(t)$ is the $(s+n)th$-order cumulant (9) of groups of operators (1).

We note that within the framework of the description of states by means of marginal density operators defined by cluster expansions over marginal correlation operators

$$F_s^{0,\epsilon}(1, \dots, s) = \sum_{\mathrm{P}:(1,\dots,s)=\cup_i X_i} \prod\nolimits_{X_i \subset \mathrm{P}} G_{|X_i|}^{0,\epsilon}(X_i), \quad s \geq 1,$$

initial states described like to sequence (15) is specified by the sequence $F^{(c)} = \left(I, F_1^{0,\epsilon}(1), \dots, \prod_{i=1}^{n} F_1^{0,\epsilon}(i), \dots\right)$, and in the case of sequence (16), the marginal density operators are represented by the following series expansions (a nonperturbative solution of the quantum BBGKY hierarchy [2]):

$$F_s(t, 1, \dots, s) = \sum_{n=0}^{\infty} \frac{1}{n!} \mathrm{Tr}_{s+1,\dots,s+n} \mathfrak{A}_{1+n}(t; \{1, \dots, s\}, s+1, \dots, s+n) \prod_{i=1}^{s+n} F_1^{0,\epsilon}(i), \quad s \geq 1,$$

where the generating operator $\mathfrak{A}_{1+n}(t)$ is the $(1+n)th$-order cumulant of groups of operators (1).

One of the possible methods to derive series expansion (13) for the marginal correlation operators lies in the substitution of the cluster expansions of groups of nonlinear operators (3) over cumulants (14) and the sequence of initial correlation operators $g(0) = \left(I, g_1^{0,\epsilon}(1), \dots, g_n^{0,\epsilon}(1, \dots, n), \dots\right)$ determined by means of the marginal correlation operators

$$g_s^{0,\epsilon}(1, \dots, s) \doteq \sum_{n=0}^{\infty} (-1)^n \frac{1}{n!} \mathrm{Tr}_{s+1,\dots,s+n} G_{s+n}^{0,\epsilon}(1, \dots, s+n), \quad s \geq 1, \tag{17}$$

into the definition of marginal correlation operators (10). Indeed, developing the generating operators of series (13) as the following cluster expansions:

$$\mathcal{G}(t; 1, \dots, s+n | f) = \sum_{\mathrm{P}:(1,\dots,s+n)=\cup_k X_k} \mathfrak{A}_{|X_1|}\left(t; X_1 | \dots \mathfrak{A}_{|X_{|\mathrm{P}|}|}\left(t; X_{|\mathrm{P}|} | f\right) \dots\right), \quad n \geq 0, \tag{18}$$

according to definition (17), we derive expressions (13). The solutions of recursive relations (18) are represented by expansions (14).

We remark that on the space $\mathfrak{L}^1(\mathcal{F}_{\mathcal{H}})$, the generating operator (14) of series expansion (13) can be represented as the $(1+n)th$-order reduced cumulant of the groups of nonlinear operators (3) of the von Neumann hierarchy [2]:

$$U_{1+n}(t;\{1,\ldots,s\},s+1,\ldots,s+n|G(0)) \doteq$$
$$\sum_{k=0}^{n}(-1)^{k}\frac{n!}{k!(n-k)!}\sum_{\mathrm{P}:(\theta(\{1,\ldots,s\}),s+1,\ldots,s+n-k)=\cup_i X_i}\mathfrak{A}_{|\mathrm{P}|}\big(t,\{X_1\},\ldots,\{X_{|\mathrm{P}|}\}\big)$$
$$\sum_{k_1=0}^{k}\frac{k!}{k_1!(k-k_1)!}\cdots\sum_{k_{|\mathrm{P}|-1}=0}^{k_{|\mathrm{P}|-2}}\frac{k_{|\mathrm{P}|-2}!}{k_{|\mathrm{P}|-1}!\big(k_{|\mathrm{P}|-2}-k_{|\mathrm{P}|-1}\big)!}G^{0,\epsilon}_{|X_1|+k-k_1}(X_1,$$
$$s+n-k+1,\ldots,s+n-k_1)\ldots G^{0,\epsilon}_{|X_{|\mathrm{P}|}|+k_{|\mathrm{P}|-1}}\big(X_{|\mathrm{P}|},s+n-k_{|\mathrm{P}|-1}+1,\ldots,s+n\big),\quad n\geq 0, \tag{19}$$

as examples, we adduce the simplest examples of reduced cumulants (19):

$$U_1(t;\{1,\ldots,s\}|G(0)) = \mathcal{G}(t;1,\ldots,s|G(0)) =$$
$$\sum_{\mathrm{P}:(1,\ldots,s)=\cup_i X_i}\mathfrak{A}_{|\mathrm{P}|}\big(t,\{X_1\},\ldots,\{X_{|\mathrm{P}|}\}\big)\prod_{X_i\subset \mathrm{P}}G^{0,\varepsilon}_{|X_i|}(X_i),$$
$$U_{1+1}(t;\{1,\ldots,s\},s+1|G(0)) = \sum_{\mathrm{P}:(1,\ldots,s+1)=\cup_i X_i}\mathfrak{A}_{|\mathrm{P}|}\big(t,\{X_1\},\ldots,\{X_{|\mathrm{P}|}\}\big)\prod_{X_i\subset \mathrm{P}}G^{0,\varepsilon}_{|X_i|}(X_i)-$$
$$\sum_{\mathrm{P}:(1,\ldots,s)=\cup_i X_i}\mathfrak{A}_{|\mathrm{P}|}\big(t,\{X_1\},\ldots,\{X_{|\mathrm{P}|}\}\big)\sum_{j=1}^{|\mathrm{P}|}G^{0,\varepsilon}_{|X_j|+1}(X_j,s+1)\prod_{\substack{X_i\subset \mathrm{P},\\ X_i\neq X_j}}G^{0,\varepsilon}_{|X_i|}(X_i).$$

We note also that a nonperturbative solution of the nonlinear quantum BBGKY hierarchy (13) or in the form of series expansions with generating operators (19) can be transformed to the perturbation (iteration) series as a result of the application of analogs of the Duhamel equation to cumulants (4) of groups of operators (1).

The following statement is true [7]. If $\max_{n\geq 1}\|G^{0,\epsilon}_n\|_{\mathfrak{L}^1(\mathcal{H}_n)} < (2e^3)^{-1}$, then in the case of bounded interaction potentials for $t\in\mathbb{R}$, a solution of the Cauchy problem of the nonlinear quantum BBGKY hierarchy (11) and (12) is determined by a sequence of marginal correlation operators represented by series expansions (13). If $G^{0,\epsilon}_n\in\mathfrak{L}^1_0(\mathcal{H}_n)\subset\mathfrak{L}^1(\mathcal{H}_n)$, it is a strong solution, and for arbitrary initial data $G^{0,\epsilon}_n\in\mathfrak{L}^1(\mathcal{H}_n)$, it is a weak solution.

3.2 A mean field asymptotic behavior of marginal correlation operators

Now we deal with a scaling asymptotic behavior of the constructed marginal correlation operators in a mean field limit in the case of initial state satisfied condition (15).

Let us observe that if $f_s\in\mathfrak{L}^1(\mathcal{H}_s)$, then for arbitrary finite time interval for an asymptotically perturbed first-order cumulant (9) of the groups of operators (1), i.e., for the strongly continuous group (1), the following equality is valid:

$$\lim_{\epsilon\to 0}\|\mathcal{G}^*_s(t,1,\ldots,s)f_s-\prod_{j=1}^{s}\mathcal{G}^*_1(t,j)f_s\|_{\mathfrak{L}^1(\mathcal{H}_s)}=0.$$

As a result of this for the $(s+n)th$-order cumulants of asymptotically perturbed groups of operators (1), the following equalities are true:

$$\lim_{\epsilon\to 0}\|\frac{1}{\epsilon^n}\mathfrak{A}_{s+n}(t,1,\ldots,s+n)f_{s+n}\|_{\mathfrak{L}^1(\mathcal{H}_{s+n})}=0,\quad s\geq 2. \tag{20}$$

We assume the existence of a mean field limit for initial marginal correlation operator (or a one-particle density operator) in the following sense:

$$\lim_{\epsilon\to 0}\|\epsilon G_1^{0,\epsilon}-g_1^0\|_{\mathfrak{L}^1(\mathcal{H})}=0. \tag{21}$$

Then, taking into account equality (20), and since the *nth* term of series expansion (16) for s-particle marginal correlation operator is determined by the $(s+n)th$-order cumulant of asymptotically perturbed groups of operators (1), we establish the property of the propagation of initial chaos (15):

$$\lim_{\epsilon\to 0}\|\epsilon^s G_s(t)\|_{\mathfrak{L}^1(\mathcal{H}_s)}=0,\quad s\geq 2. \tag{22}$$

If for the initial marginal correlation operator equality (21) holds, then in the case of $s=1$ for series expansion (16), the following equality is true:

$$\lim_{\epsilon\to 0}\|\epsilon G_1(t)-g_1(t)\|_{\mathfrak{L}^1(\mathcal{H})}=0,$$

where for arbitrary finite time interval, the limit one-particle marginal correlation operator $g_1(t,1)$ is given by the norm convergent series on the space $\mathfrak{L}^1(\mathcal{H})$

$$\begin{aligned} g_1(t,1)=\sum_{n=0}^{\infty}\int_0^t dt_1\ldots\int_0^{t_{n-1}}dt_n\,\mathrm{Tr}_{2,\ldots,n+1}\mathcal{G}_1^*(t-t_1,1)\mathcal{N}_{\mathrm{int}}^*(1,2)\prod_{j_1=1}^{2}\mathcal{G}_1^*(t_1-t_2,j_1)\ldots \\ \prod_{i_n=1}^{n}\mathcal{G}_1^*(t_n-t_n,i_n)\sum_{k_n=1}^{n}\mathcal{N}_{\mathrm{int}}^*(k_n,n+1)\prod_{j_n=1}^{n+1}\mathcal{G}_1^*(t_n,j_n)\prod_{i=1}^{n+1}g_1^0(i). \end{aligned} \tag{23}$$

In series expansion (23), the operator $\mathcal{N}_{\mathrm{int}}^*(j_1,j_2)$ is defined according to formula (2), and the group of operators $\mathcal{G}_1^*(t)$ is defined by (1). For bounded interaction potential, series (23) is norm convergent on the space $\mathfrak{L}^1(\mathcal{H})$ under the condition that $t<t_0\equiv\left(2\|\Phi\|_{\mathfrak{L}(\mathcal{H}_2)}\|g_1^0\|_{\mathfrak{L}^1(\mathcal{H})}\right)^{-1}$.

As a result of differentiation in the sense of the norm convergence of the space $\mathfrak{L}^1(\mathcal{H})$ by the time variable of the operator represented by series expansion (23), we conclude that limit one-particle marginal correlation operator (23) is governed by the Cauchy problem of the quantum Vlasov kinetic equation:

$$\frac{\partial}{\partial t}g_1(t,1)=\mathcal{N}^*(1)g_1(t,1)+\mathrm{Tr}_2\mathcal{N}_{\mathrm{int}}^*(1,2)g_1(t,1)g_1(t,2), \tag{24}$$

$$g_1(t)\big|_{t=0}=g_1^0. \tag{25}$$

Then for pure states we derive the Hartree equation [2], indeed, in terms of the kernel $g_1(t,q;q')=\psi(t,q)\psi(t,q')$ of operator (23), describing a pure-state, quantum kinetic equation (24) is reduced to the Hartree equation

$$i\frac{\partial}{\partial t}\psi(t,q)=-\frac{1}{2}\Delta_q\psi(t,q)+\int dq'\Phi(q-q')|\psi(t,q')|^2\psi(t,q),$$

where the function Φ is a two-body interaction potential.

We note that in the case of pure states, kinetic equation (24) can be reduced to the nonlinear Schrödinger equation [12] or to the Gross-Pitaevskii kinetic equation [13].

4. The description of processes of a creation and a propagation of correlations by means of kinetic equations

In this section we consider mathematical problems concerning the description of processes of creation and propagation of correlations within framework of the state of typical particle of quantum systems of many particles; in other words, an approach to the description of evolution of correlations by means of quantum kinetic equations is developing.

4.1 Marginal correlation functionals of the state

Further we shall consider the case of initial states specified by a one-particle marginal density operator with correlations, namely, initial states specified by the following sequence of marginal correlation operators:

$$G^{(c)} = \left(I, G_1^{0,\epsilon}(1), g_2^{\epsilon}(1,2)\prod_{i=1}^{2} G_1^{0,\epsilon}(i), ..., g_n^{\epsilon}(1,...,n)\prod_{i=1}^{n} G_1^{0,\epsilon}(i), ...\right), \qquad (26)$$

where the operators $g_n^{\epsilon}(1,...,n) \equiv g_n^{\epsilon} \in \mathfrak{L}_0^1(\mathcal{H}_n), n \geq 2$ specified the initial correlations. We remark that such assumption about initial states is intrinsic for the kinetic description of many-particle systems. On the other hand, initial data (26) is typical for the condensed states of large quantum systems of particle, for example, the equilibrium state of the Bose condensate satisfies the weakening of correlation condition with the correlations which characterize the condensed state [1].

For initial states specified in terms of a one-particle density operator and correlation operators (26), the evolution of states given within the framework of the sequence $G(t) = (I, G_1(t), ..., G_s(t), ...)$ of marginal correlation operators (13) can be described by means of the sequence $G(t|G_1(t)) = (I, G_1(t), G_2(t|G_1(t)), ..., G_s(t|G_1(t)), ...)$ of marginal correlation functionals: $G_s(t,1,...,s|G_1(t)), s \geq 2$, with respect to the one-particle correlation operator $G_1(t)$ governed by the kinetic equation [8].

In the case under consideration, the marginal correlation functionals $G_s(t|G_1(t)), s \geq 2$ are defined with respect to the one-particle (marginal) density operator

$$G_1(t,1) = \sum_{n=0}^{\infty}\frac{1}{n!}\mathrm{Tr}_{2,...,1+n}\mathfrak{A}_{1+n}(t,1,...,n+1)\sum_{\mathrm{P}\,:\,(1,...,n+1)=\cup_i X_i}\prod_{X_i\subset \mathrm{P}} g_{|X_i|}^{\epsilon}(X_i)\prod_{i=1}^{n+1} G_1^{0,\epsilon}(i), \qquad (27)$$

where the generating operator $\mathfrak{A}_{1+n}(t)$ is the $(1+n)-th$-order cumulant (4) of the groups of operators (1), and these functionals are represented by the series expansions:

$$G_s(t,1,...,s|G_1(t)) = \sum_{n=0}^{\infty}\frac{1}{n!}\mathrm{Tr}_{s+1,...,s+n}\mathfrak{G}_{s+n}(t,\theta(\{1,...,s\}),s+1,...,s+n)\prod_{i=1}^{s+n} G_1(t,i), \quad s \geq 2, \qquad (28)$$

where the $(s+n)th$-order generating operator $\mathfrak{G}_{s+n}(t), n \geq 0$ of this series is determined by the following expansion:

$$\mathfrak{G}_{s+n}(t,\theta(\{1,\ldots,s\}),s+1,\ldots,s+n) = n!\sum_{k=0}^{n}(-1)^k\sum_{n_1=1}^{n}\ldots\sum_{n_k=1}^{n-n_1-\ldots-n_{k-1}}\frac{1}{(n-n_1-\ldots-n_k)!}\times \breve{\mathfrak{A}}_{s+n-n_1-\ldots-n_k}(t,\theta(\{1,\ldots,s\}),s+1,\ldots,s+n-n_1-\ldots-n_k)\times \prod_{j=1}^{k}\sum_{\substack{\mathrm{D}_j:Z_j=\cup_{l_j}X_{l_j},\\ |\mathrm{D}_j|\leq s+n-n_1-\ldots-n_j}}\frac{1}{|\mathrm{D}_j|!}\sum_{i_1\neq\ldots\neq i_{|\mathrm{D}_j|}=1}^{s+n-n_1-\ldots-n_j}\prod_{X_{l_j}\subset\mathrm{D}_j}\frac{1}{|X_{l_j}|!}\breve{\mathfrak{A}}_{1+|X_{l_j}|}\left(t,i_{l_j},X_{l_j}\right). \tag{29}$$

In formula (29) the sum of all possible dissections [18] of the linearly ordered set $Z_j\equiv\left(s+n-n_1-\ldots-n_j+1,\ldots,s+n-n_1-\ldots-n_{j-1}\right)$ on no more than $s+n-n_1-\ldots-n_j$ linearly ordered subsets is denoted by $\sum_{\mathrm{D}_j:Z_j=\cup_{l_j}X_{l_j}}$, and the $(s+n)th$-order scattering cumulant is defined by the formula

$$\breve{\mathfrak{A}}_{s+n}(t,\theta(\{1,\ldots,s\}),s+1,\ldots,s+n)\doteq\mathfrak{A}_{s+n}(t,1,\ldots,s+n)g^{\epsilon}_{s+n}(1,\ldots,s+n)\prod_{i=1}^{s+n}\mathfrak{A}_1^{-1}(t,i),$$

where the operator $g^{\epsilon}_{s+n}(1,\ldots,s+n)$ is specified initial correlations (26) and notations accepted above were used. We adduce simplest examples of generating operators (29):

$$\mathfrak{G}_s(t,\theta(\{1,\ldots,s\}))=\breve{\mathfrak{A}}_s(t,\theta(\{1,\ldots,s\}))=\mathfrak{A}_s(t,1,\ldots,s))g^{\epsilon}_s(1,\ldots,s)\prod_{i=1}^{s}\mathfrak{A}_1^{-1}(t,i),$$

$$\mathfrak{G}_{s+1}(t,\theta(\{1,\ldots,s\}),s+1)=\mathfrak{A}_{s+1}(t,1,\ldots,s+1)g^{\epsilon}_{s+1}(1,\ldots,s+1)\prod_{i=1}^{s+1}\mathfrak{A}_1^{-1}(t,i)-$$

$$\mathfrak{A}_s(t,1,\ldots,s)g^{\epsilon}_s(1,\ldots,s)\prod_{i=1}^{s}\mathfrak{A}_1^{-1}(t,i)\sum_{j=1}^{s}\mathfrak{A}_2(t,j,s+1)g^{\epsilon}_2(j,s+1)\mathfrak{A}_1^{-1}(t,j)\mathfrak{A}_1^{-1}(t,s+1).$$

A method of the construction of marginal correlation functionals (28) is based on the application of kinetic cluster expansions [2] to the generating operators of series (13). If $\|G_1(t)\|_{\mathfrak{L}^1(\mathcal{H})}<e^{-(3s+2)}$, then for arbitrary $t\in\mathbb{R}$ series expansion (28) converges in the norm of the space $\mathfrak{L}^1(\mathcal{H}_s)$.

We emphasize that marginal correlation functionals (28) describe all the possible correlations generated by dynamics of large quantum particle system with initial correlations by means of a one-particle density operator.

4.2 The generalized quantum kinetic equation with initial correlations

Now we establish the evolution equation for one-particle (marginal) density operator (27). As a result of the differentiation over time variable of the operator represented by series expansion (27) in the sense of the norm convergence of the space $\mathfrak{L}^1(\mathcal{H})$, then due to the application of the kinetic cluster expansions [19] to the generating operators of obtained series expansion, for one-particle density operator (27), we derive the following identity:

$$\frac{\partial}{\partial t}G_1(t,1) = \mathcal{N}^*(1)G_1(t,1) + \epsilon \mathrm{Tr}_2\mathcal{N}^*_{\mathrm{int}}(1,2)G_1(t,1)G_1(t,2) \\ + \epsilon \mathrm{Tr}_2\mathcal{N}^*_{\mathrm{int}}(1,2)G_2(t,1,2|G_1(t)), \tag{30}$$

where the second part of the collision integral in equality (30) is determined in terms of the marginal correlation functional represented by series expansions (28) in the case of $s = 2$. This identity we treat as the quantum kinetic equation, and we refer to this evolution equation as the generalized quantum kinetic equation with initial correlations.

We emphasize that the coefficients in an expansion of the collision integral of the non-Markovian kinetic equation (30) are determined by the operators specified initial correlations (26).

On the space $\mathfrak{L}^1(\mathcal{H})$ for the Cauchy problem of the established generalized quantum kinetic equation with initial correlations, the following statement is true [19]. If $\|G_1^{0,\epsilon}\|_{\mathfrak{L}^1(\mathcal{H})} < (e(1+e^9))^{-1}$, a global in time solution of the Cauchy problem of kinetic equation (30) is determined by series expansion (27). For initial data $G_1^{0,\epsilon} \in \mathfrak{L}^1_0(\mathcal{H})$, it is a strong solution, and for an arbitrary initial data, it is a weak solution.

The proof of this existence statement is similar to the proof in the case of the generalized quantum kinetic equation given in [18].

4.3 On a propagation of initial correlations in a mean field limit

Further we establish the mean field asymptotic behavior of constructed marginal correlation functionals (28) in the case of initial states specified by the one-particle density operator with correlations (26).

We assume the existence of a mean field limit of an initial one-particle density operator in sense (21) and for initial correlation operators as follows:

$$\lim_{\epsilon\to 0} \|g_n^\epsilon - g_n\|_{\mathfrak{L}^1(\mathcal{H}_n)} = 0, \quad n \geq 2. \tag{31}$$

Then in consequence of the validity of equalities (20) for one-particle density operator (27), the following statement is true [8]. If conditions (21) and (31) hold, then for series expansion (27) the equality holds:

$$\lim_{\epsilon\to 0} \|\epsilon G_1(t) - g_1(t)\|_{\mathfrak{L}^1(\mathcal{H})} = 0,$$

where for finite time interval, the limit one-particle density operator $g_1(t)$ is represented by the following norm convergent series on the space $\mathfrak{L}^1(\mathcal{H})$:

$$g_1(t,1) = \sum_{n=0}^{\infty} \int_0^t dt_1 \ldots \int_0^{t_{n-1}} dt_n \, \mathrm{Tr}_{2,\ldots,n+1}\mathcal{G}_1^*(t-t_1,1)\mathcal{N}^*_{\mathrm{int}}(1,2)\prod_{j_1=1}^{2}\mathcal{G}_1^*(t_1-t_2,j_1)\ldots \\ \prod_{i_n=1}^{n}\mathcal{G}_1^*(t_n-t_n,i_n)\sum_{k_n=1}^{n}\mathcal{N}^*_{\mathrm{int}}(k_n,n+1)\prod_{j_n=1}^{n+1}\mathcal{G}_1^*(t_n,j_n) \sum_{\mathrm{P}:(1,\ldots,n+1)=\cup_i X_i} \\ \prod_{X_i\subset \mathrm{P}} g_{|X_i|}(X_i)\prod_{i=1}^{n+1} g_1^0(i). \tag{32}$$

In series expansion (32), the operator $\mathcal{N}^*_{\mathrm{int}}(j_1,j_2)$ is defined according to formula (2), and the group of operators $\mathcal{G}_1^*(t)$ is defined by (1). For bounded interaction potentials series (32) is norm convergent on the space $\mathfrak{L}^1(\mathcal{H})$ under the condition that $t < t_0 \equiv \left(2\|\Phi\|_{\mathfrak{L}(\mathcal{H}_2)}\|g_1^0\|_{\mathfrak{L}^1(\mathcal{H})}\right)^{-1}$.

For marginal correlation functionals (28), the following statement is true [8]. Under conditions (21) and (31) on initial state (26), there exists a mean field limit of marginal correlation functionals (28) in the following sense:

$$\lim_{\epsilon\to 0}\|\epsilon^s G_s(t,1,\dots,s|G_1(t)) - g_s(t,1,\dots,s|g_1(t))\|_{\mathfrak{L}^1(\mathcal{H}_s)} = 0, \quad s\geq 2,$$

where the limit marginal correlation functionals $g_s(t|g_1(t)), s\geq 2,$ are represented by the expansions

$$g_s(t,1,\dots,s|g_1(t)) = \prod_{i_1=1}^{s}\mathcal{G}_1^*(t,i_1)g_s(1,\dots,s)\prod_{i_2=1}^{s}\left(\mathcal{G}_1^*\right)^{-1}(t,i_2)\prod_{j=1}^{s}g_1(t,j), \tag{33}$$

and, respectively, the limit one-particle density operator $g_1(t)$ is represented by series expansion (32).

The proof of these statements is based on the validity of equality (20) for cumulants of asymptotically perturbed groups of operators (1) and the explicit structure of the generating operators of series expansions (28) of marginal correlation functionals and of series expansion (27).

We remark that limit marginal correlation functionals (32) and (33) are a solution of the Cauchy problem of the quantum Vlasov hierarchy of nonlinear evolution equations [6], which describes a mean field asymptotic behavior of marginal correlation operators in the case of arbitrary initial states, namely,

$$\frac{\partial}{\partial t}g_s(t,1,\dots,s) = \sum_{i=1}^{s}\mathcal{N}^*(i)g_s(t,1,\dots,s) + \mathrm{Tr}_{s+1}\sum_{i=1}^{s}\mathcal{N}_{\mathrm{int}}^*(i,s+1)$$
$$\Big(g_{s+1}(t,1,\dots,s+1) + \sum_{\substack{\mathrm{P}:\,(1,\dots,s+1)=X_1\cup X_2,\\ i\in X_1;\, s+1\in X_2}} g_{|X_1|}(t,X_1)g_{|X_2|}(t,X_2)\Big),$$
$$g_s(t)\big|_{t=0} = g_s^0, \quad s\geq 1,$$

where we used notations similar to accepted above.

It should be noted that limit marginal correlation functionals (33) describe the process of the evolution of correlations of large quantum particle systems by means of a one-particle density operator in a mean field approximation.

Similar to the derivation of kinetic equation (30), we establish that the one-particle density operator represented by series expansion (32) is a solution of the Cauchy problem of the Vlasov-type quantum kinetic equation with initial correlations:

$$\frac{\partial}{\partial t}g_1(t,1) = \mathcal{N}^*(1)g_1(t,1) +$$
$$\mathrm{Tr}_2\mathcal{N}_{\mathrm{int}}^*(1,2)\prod_{i_1=1}^{2}\mathcal{G}_1^*(t,i_1)\left(g_2(1,2)+I\right)\prod_{i_2=1}^{2}\left(\mathcal{G}_1^*\right)^{-1}(t,i_2)g_1(t,1)g_1(t,2), \tag{34}$$

$$g_1(t)\big|_{t=0} = g_1^0, \tag{35}$$

and consequently, for pure states we derive the Hartree-type equation with initial correlations. We point out that Eq. (34) is the non-Markovian quantum kinetic equation.

Thus, we established that a mean field behavior of processes of the creation of correlations and the propagation of initial correlations in large quantum particle systems are governed by kinetic equation (34).

Moreover, in the case under consideration, the processes of the creation of correlations generated by dynamics of many-particle systems and the propagation of initial correlations are described by the constructed marginal functionals of the state (28) governed by the non-Markovian generalized kinetic equation with initial correlations (26).

5. Conclusion

In this chapter the process of a creation and a propagation of correlations in quantum many-particle systems has been described by means of the Cauchy problem of the quantum BBGKY hierarchy of nonlinear equations (11) and (12). A nonperturbative solution for a sequence of marginal correlation operators is represented in the form of series (13) the generating operator of every term of which are corresponding-order cumulant (14) of groups of nonlinear operators (3). In the case of initial state specified by a sequence of the marginal correlation operators that satisfy chaos property (15), the correlations generated by dynamics of large quantum particle system (16) are completely determined by the corresponding-order cumulants (4) of groups of operators (1). The obtained results can be extended to large quantum systems of bosons and fermions like in paper [6].

In the case of initial state satisfied condition (15), a mean field asymptotic behavior of the processes of a creation and a propagation of correlations was described. It was directly proven the property called the propagation of initial chaos (22), which underlies in mathematical derivation of effective evolution equations of systems of infinitely many particles [16].

The problem of the rigorous description of collective behavior of quantum many-particle systems by means of a one-particle (marginal) correlation operator that is a solution of the generalized quantum kinetic equation [18] with initial correlations [19], for instance, the initial correlations, characterizing the condensed states [1], or initial correlations that influence on ultrafast relaxation processes in plasmas [4] has been also considered.

In particular, such an approach to the derivation of the Vlasov-type quantum kinetic equation with initial correlations (34) from underlying dynamics governed by the generalized quantum kinetic equation with initial correlations (30) enables to construct the higher-order corrections to the mean field evolution of large quantum systems of particle.

We note that in paper [20] other approach to the description of the propagation of initial correlations of large quantum particle systems in a mean field limit was developed, namely, the process of the propagation of initial correlations was described within the framework of the evolution of marginal observables governed by the dual BBGKY hierarchy [2, 21].

DOI: http://dx.doi.org/10.5772/intechopen.82836

Author details

Viktor I. Gerasimenko
Institute of Mathematics of the NAS of Ukraine, Kyiv, Ukraine

*Address all correspondence to: gerasym@imath.kiev.ua

References

[1] Bogolyubov MM. Lectures on Quantum Statistics. Problems of Statistical Mechanics of Quantum Systems. Kyiv: Rad. Shkola; 1949 (in Ukrainian)

[2] Gerasimenko VI. Hierarchies of quantum evolution equations and dynamics of many-particle correlations. In: Statistical Mechanics and Random Walks: Principles, Processes and Applications. N.Y.: Nova Science Publ., Inc.; 2012. p. 233

[3] Prigogine I. Non-Equilibrium Statistical Mechanics. New York: Wiley; 1962

[4] Bonitz M, Henning C, Block D. Complex plasmas: A laboratory for strong correlations. Reports on Progress in Physics. 2010;**73**:066501 (29 p)

[5] Gerasimenko VI, Shtyk VO. Evolution of correlations of quantum many-particle systems. Journal of Statistical Mechanics: Theory and Experiment. 2008;**3**:P03007 (24 p)

[6] Gerasimenko VI, Polishchuk DO. Dynamics of correlations of Bose and Fermi particles. Mathematical Models and Methods in Applied Sciences. 2011; **34**(1):76

[7] Gerasimenko VI. Evolution of correlation operators of large quantum particle systems. Methods of Functional Analysis and Topology. 2017;**23**(2):123

[8] Gerasimenko VI. On the description of quantum correlations by means of a one-particle density operator. Proc IM NASU. 2017;**14**(1):116

[9] Spohn H. Kinetic equations from Hamiltonian dynamics. Reviews of Modern Physics. 1980;**53**:600-640

[10] Benedetto D, Castella F, Esposito R, Pulvirenti M. A short review on the derivation of the nonlinear quantum Boltzmann equations. Communications in Mathematical Sciences. 2007;**5**:55-71

[11] Pezzotti F, Pulvirenti M. Mean-field limit and semiclassical expansion of quantum particle system. Annales Henri Poincaré. 2009;**10**:145-187

[12] Erdös L, Schlein B, Yau H-T. Derivation of the cubic nonlinear Schrödinger equation from quantum dynamics of many-body systems. Inventiones Mathematicae. 2007; **167**(3):515-614

[13] Erdös L, Schlein B, Yau H-T. Derivation of the Gross–Pitaevskii equation for the dynamics of Bose–Einstein condensate. Annals of Mathematics. 2010;**172**:291-370

[14] Boccato C, Cenatiempo S, Schlein B. Quantum many-body fluctuations around nonlinear Schrödinger dynamics. Annales Henri Poincaré. 2017;**18**(1):113-191

[15] Porta M, Rademacher S, Saffirio C, Schlein B. Mean field evolution of fermions with coulomb interaction. Journal of Statistical Physics. 2017; **166**(6):1345-1364

[16] Golse F. On the dynamics of large particle systems in the mean field limit. Macroscopic and large scale phenomena: Coarse graining, mean field limits and ergodicity. Lecture Notes in Applied Mathematics and Mechanics. Springer. 2016;**3**:1-144

[17] Golse F, Mouhot C, Paul T. On the mean-field and classical limits of quantum mechanics. Communications in Mathematical Physics. 2016;**343**: 165-205

[18] Gerasimenko VI, Tsvir ZA. A description of the evolution of quantum states by means of the kinetic equation. Journal of Physics A: Mathematical and Theoretical. 2010;**43**(48):485203

[19] Gerasimenko VI, Tsvir ZA. On quantum kinetic equations of many-particle systems in condensed states. Physica A: Statistical Mechanics and its Applications. 2012;**391**(24):6362

[20] Gerasimenko VI. New approach to derivation of quantum kinetic equations with initial correlations. Carpathian Mathematical Publications. 2015;7:38-48

[21] Gerasimenko VI. Approaches to derivation of quantum kinetic equations. Ukrainian Journal de Physique. 2009;**54**(8–9):834

Chapter 4

Recent Progresses in *Ab Initio* Electronic Structure Calculation toward Understandings of Functional Mechanisms of Biological Macromolecular Systems

Jiyoung Kang, Takuya Sumi and Masaru Tateno

Abstract

In this chapter, we present recent advances of theoretical analyses toward understandings of functional mechanisms of biological macromolecular systems, employing *ab initio* electronic structure calculations. Two distinct types of triggers to invoke dramatic rearrangements of electronic structures in the reaction centers are revealed by full *ab initio* quantum mechanics (QM) calculations (first example) and hybrid *ab initio* QM/molecular mechanics (MM) molecular dynamics (MD) calculations (second example). First, we demonstrate dramatic rearrangements of molecular orbitals (MOs) induced by binding of a hydroxyl ion (OH^-) to the [4Fe-3S] cluster found in hydrogenases, which catalyzes both dissociation and production of dihydrogen (H_2). This induces the significant delocalization of the LUMO, resulting in formation of electron transfer pathways required for the catalysis. Thus, in organisms, just a tiny species (e.g. OH^- ligand) can play a key role for the biological functions. Second, we indicate dynamical rearrangements of MOs occurring in the enzymatic reactions of RNA-protein complexes. As the catalysis proceeds, the reactive MOs, which do not belong to the frontier orbitals in the initial stages of the reaction, are dramatically reconstituted in the hybrid *ab initio* QM/MM MD simulations, resulting in the frontier orbitals, which is a feature characteristic to biological macromolecular systems.

Keywords: density functional theory (DFT), electron delocalization, aminoacyl-tRNA synthetase (aaRS), transfer RNA (tRNA), hybrid ribozyme/protein catalyst, catalysis, enzyme, molecular evolution

1. Introduction

Why are the theoretical analyses employing *ab initio* quantum mechanics (QM) calculations required to understand biological systems? In an organism, so many catalytic reactions are present, for example, transcription, DNA replication and repair, protein biosynthesis, respiration, photosynthesis, and synthesis and degradation of biological compounds (metabolites) such as amino acids, nucleotide, and lipid. In order to understand the mechanisms of biological functions, analyses of the electronic structure changes for the catalytic reactions are essential.

Up to date, QM calculations have been employed to understand many biochemical reactions, although the system sizes of such macromolecular systems are huge. In this section, we will briefly introduce several substantial issues in QM methods that have frequently been employed in analyses of biological systems. From biochemical and biophysical points of view, these descriptions are also relevant to the construction of the QM models, spin assignments, selection of QM/MM methods, the QM calculation methods, basis sets, and so on.

1.1 Construction of QM model system

To obtain precise geometric and electronic structures employing the QM calculations, high-quality three-dimensional (3D) structures are indispensable. In most studies of biological macromolecular systems, the initial 3D structures for the theoretical analyses are retrieved from Protein Data Bank (PDB) website (https://www.rc sb.org/), which provides 3D structures of biological macromolecules analyzed employing X-ray crystallography, nuclear magnetic resonance (NMR) spectroscopy, and electron microscope (EM) experiments. Currently, the PDB site contains more than 129,300 X-ray structures, 12,300 NMR structures, and 2400 EM structures.

Although state-of-the-art methodologies, such as the X-ray free electron laser (XFEL) and cryo-EM, provide high-quality 3D structures, the resolution of most experimental structures is still insufficient to observe hydrogen atoms. Indeed, only 0.5% of X-ray and EM structures are under 1.0 Å resolutions, and 80.3% are in the range of 1.4–2.8 Å resolutions. Thus, one needs to attach the hydrogen atoms in chemically appropriate manners. This is also an important issue, and so we need enough time to carefully identify the appropriate configurations for attachments of hydrogen atoms.

Since the computation costs of QM calculations are too large to include the entire biological macromolecular systems, QM models are usually extracted and thereby include the numbers of atoms in the ranges of 50–100 atoms. The truncated boundary carbons of the QM models are usually capped by the methyl group. Other crucial moieties in the systems, which can significantly affect geometric and electronic structures of the active centers, such as ligands of the transition metal binding sites and hydrogen-bonded waters, should also be included in the extracted QM models. Notably, to overcome the increase of computational costs by including large environmental moieties into the QM models such as the bulk water molecules as the solvent, hybrid quantum mechanics and molecular mechanics (MM) (i.e., classical mechanics) schemes have been developed up to date (the hybrid QM/MM calculation method is discussed in Section 1.3).

1.2 Spin assignments of the system

As mentioned, we require 3D structures of the calculation models, basis sets, the (total) charges, spin multiplicities, and initial wave functions of the systems, to perform the QM calculations. Since the QM calculations may also suffer from the nonlinear, initial guess, and local minimum problems, starting from appropriate initial 3D structures and wave functions is very important to obtain the correct states. In particular, the total spin of the system including multiple transition metals must be explored by carefully providing multiple combinations of spin assignments and thereby should be determined by identifying the spin state with the optimal total energy value among all of those spin states.

To provide the total spin assignments, it is convenient to divide the systems into some fragmental moieties, such as a transition metal and its coordinated ligands as a subsystem, which leads to spin assignments for a part of the QM models. Although this could be helpful to get a reasonable solution, there is no warranty of the convergence and acquisition of the correct solution. In Section 2, we discuss more details of this issue.

1.3 Hybrid functional in density functional theory (DFT) calculation

Hybrid functional approximation was introduced by Becke [1] in 1993 and has become one of the most popular computational approaches. By incorporating a portion of the exact exchange energy originated from the Hartree-Fock theory coupled with the exchange-correlation energies, the hybrid functional approach improves molecular properties that poorly described with simple *ab initio* functionals, such as bond lengths, vibration frequencies, and atomization energies [2]. In general, a hybrid functional is described as a linear combination of the Hartree-Fock exact exchange energy functional and the exchange-correlation density functionals; their weights are determined by a fitting procedure such as to reproduce experimental or highly advanced calculated thermochemical data.

Becke proposed the following exchange-correlation approximation [1],

$$E_{XC} = E_{XC}^{LSDA} + a_0\left(E_X^{exact} - E_X^{LSDA}\right) + a_X\left(E_X^{B88} - E_X^{LSDA}\right) + a_C\left(E_X^{PW91} - E_X^{LSDA}\right), \quad (1)$$

where a_0, a_x, and a_c are the parameters determined by the fitting procedure (i.e., $a_0 = 0.20$, $a_x = 0.72$, and $a_c = 0.81$). Herein, subscripts X and C represent the exchange and correlation, respectively, E_{XC}^{LSDA} represents the exchange-correlation energy within the framework of the local spin-density approximation (LSDA), E_X^{exact} is the exact exchange energy, and E_X^{B88} and E_X^{PW91} are the gradient corrections for the exchange by Becke and Perdew-Wang, respectively.

The B3LYP functional, which has been one of the most widely used functionals in molecular quantum calculation fields, employs the nonlocal correlation provided by the LYP expression (Lee-Yang-Parr) and the Becke88 exchange functional [3], and VWN local-density approximation that was constructed by Volsko, Wilk, and Nusair (VWN) [4]. Thus, the general formula of the hybrid functionals can be written as follows (this is exploited in Gaussian software; http://gaussian.com/dft/),

$$E_{XC} = P_2E_X^{HF} + P_1\left(P_4E_X^{Slater} + P_3\Delta E_X^{non-local}\right) + P_6E_C^{local} + P_5E_C^{non-local}. \quad (2)$$

Various combinations of nonlocal exchange functionals and correlation functionals can be employed here. In the B3LYP functional, the Becke's three-parameter functional and the LYP correlation functional were combined with the following parameters,

$$P_1 = 1, P_2 = 0.2, P_3 = 0.72, P_4 = 0.8, P_5 = 0.81, P_6 = 1. \quad (3)$$

In an attempt to improve the B3LYP functional, the long-range correlations are incorporated into the cam-B3LYP [5] and LC-BLYP [6] functionals. However, to the best of our experiences on applying to biological macromolecular systems, selection of functionals is not so simple, and thus, careful investigations based on computational trials are required to determine them. In the following two examples described in this chapter, such examinations were actually performed, and thereby, the B3LYP functional was adopted.

1.4 Hybrid QM/MM calculation scheme

In 1976, Warshel and Levitt [7] developed a QM/MM method, in which QM calculation is combined with classical mechanics calculation, to obtain the electronic structures of the QM region with consideration of the environmental effects, such as protein, membrane, and solvent water molecules. In this strategy, the QM calculation is adopted to the active site (QM region), and for the remainder of the system, the MM calculation is adopted (MM regions) (**Figure 1**).

Great progresses have been achieved up to date for improvement of QM/MM calculation algorithms and their applications to biological systems [8–21]. Importance of the environments has been reported from many QM/MM studies. For example, polarization from MM region affects both electronic structure and geometric structure [22]. Recently, we reported that in huge biological macromolecular systems such as complexes of aminoacyl-tRNA synthetases (aaRSs) and their cognate tRNAs, dynamical, geometrical changes induced dramatical rearrangements of the electronic structures in the catalytic sites, which thus generated the productive states in the reactions [23–25] (see Section 3). By contrast, in Section 2, full (*ab initio*) QM calculations were solely employed for the analysis of the electronic structures of a transition metal cluster found in proteins, since in most cases, such a structure is buried in protein environments.

1.5 Energy expression for QM/MM calculation

In the framework of QM/MM methodology, an entire system is divided into two regions: QM region, which is described by quantum mechanics principles, and MM region, which is described by molecular mechanics (i.e., classical mechanics). Due

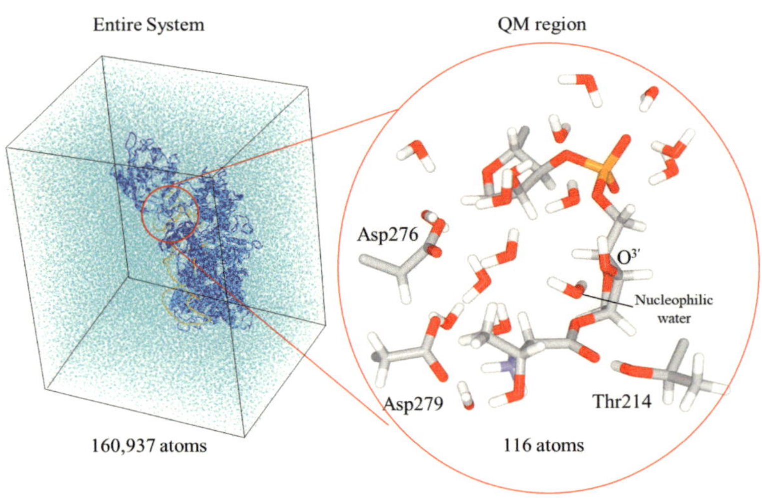

Figure 1.
Hybrid QM/MM modeling of a biological macromolecular system (i.e., valyl-tRNA synthetase (ValRS) in complex with the cognate tRNA; see Section 3).

to the presence of the interactions between QM and MM regions, the total energy of the entire system can be formally written as follows:

$$\mathrm{E(ES)} = \mathrm{E(QM)} + \mathrm{E(MM)} + \mathrm{E(QM/MM)} \tag{4}$$

The inclusion of the energy term of QM-MM interactions, E(QM/MM), enables a more realistic description of the system, compared with isolated QM calculations. In terms of the treatment of the electrostatic interaction between the QM and MM regions, QM/MM methodologies are divided into two groups, subtractive and additive schemes. Herein, we discuss the advantages and disadvantages of the QM/MM methodologies in the comparison of these two schemes.

Subtractive schemes consist of the three steps as follows: (1) an MM calculation on the entire system, (2) a QM calculation on the QM region, and (3) an MM calculation on the QM region. Then, QM/MM energy of the entire system can be formulated as follows:

$$E^{\mathrm{sub}}(\mathrm{ES}) = E_{\mathrm{MM}}(\mathrm{ES}) + E_{\mathrm{QM}}(\mathrm{QM}) - E_{\mathrm{MM}}(\mathrm{QM}) \tag{5}$$

The subscript indicates the type of calculation (QM or MM calculation), and the region on which the calculation is performed is described in parentheses. An advantage of the subtractive schemes is simplicity. Explicit descriptions of interactions between QM and MM regions are not required. In addition, artifacts that might be caused by using link atom schemes to cap the truncated bonds at the QM-MM boundary (described below) can be avoided. On the other hand, disadvantages of the subtractive schemes are the following. (1) Force fields are required for describing the QM region that often includes ligands and intermediate structures of enzymatic reactions; in general, reliable force fields of the molecules are not prepared, and additional QM calculations should be carried out for the parameterization every time a new system is studied. (2) The electrostatic interactions between the QM and MM regions are described at molecular mechanics level; that is, the interactions are calculated by the Coulomb interactions between fixed atomic charges in the QM and MM regions. Such descriptions cannot represent polarization of the QM region induced by the environment surrounding the QM region.

On the other hand, the additive schemes can take into account the polarization effects. The energy expression for the additive schemes is given in Eq. (6):

$$E^{\mathrm{add}}(\mathrm{ES}) = E_{\mathrm{MM}}(\mathrm{MM}) + E_{\mathrm{QM}}(\mathrm{QM}) + E_{\mathrm{QM/MM}}(\mathrm{QM}, \mathrm{MM}) \tag{6}$$

A characteristic feature of this scheme is the presence of the energy term with respect to the interactions between QM and MM regions, described as follows:

$$\begin{aligned} E_{\mathrm{QM/MM}}(\mathrm{QM}, \mathrm{MM}) &= E^{\mathrm{elec}}_{\mathrm{QM/MM}}(\mathrm{QM}, \mathrm{MM}) \\ &\quad + E^{\mathrm{bonded}}_{\mathrm{QM/MM}}(\mathrm{QM}, \mathrm{MM}) + E^{\mathrm{vdW}}_{\mathrm{QM/MM}}(\mathrm{QM}, \mathrm{MM}) \end{aligned} \tag{7}$$

To calculate electrostatic interactions, that is, the first term in the left side of Eq. (7), one-electron integrals in the QM Hamiltonian incorporating MM partial charges can be used.

$$\hat{H}^{\mathrm{elec}}_{\mathrm{QM/MM}}(\mathrm{QM}, \mathrm{MM}) = -\sum_i^N \sum_j^M \frac{q_j}{|\mathbf{r}_i - \mathbf{R}_j|} + \sum_k^L \sum_j^M \frac{q_j Q_k}{|\mathbf{R}_k - \mathbf{R}_j|} \tag{8}$$

The symbols q_j are the MM partial charges located at R_j. Q_k are the nuclear charges of the QM atoms at R_k, and r_i represents electron positions. N, M, and L represent number of electrons, MM atoms to be incorporated into the one-electron integrals, and QM atoms, respectively. Using the additive scheme, the electronic structures of the QM region are affected by the charge distribution of the environment. An advanced approach is to consider polarization of the MM region by the QM region (i.e., to allow the partial charges to be changed according to changes in the electronic structure of the QM region). However, polarizable force fields with broader applications have not yet emerged, while many efforts to account for the polarization effects were made [26, 27].

2. *Ab initio* QM analysis of electron transfer (ET) mechanisms in hydrogenase

2.1 [NiFe] hydrogenase

Hydrogenase is an enzyme that can catalyze dihydrogen (H_2) to water molecules and its reverse process [28]. Due to the reversible oxidation properties in the H_2 catalysis, hydrogenase has been focused in biotechnological devices, such as generation of H_2 from solar energy [29]. However, most [NiFe] hydrogenases, which are classified as standard hydrogenase, are sensitive to the explosion of the O_2; that is, their activities decrease in an aerobic condition. By contrast, some hydrogenases preserve their activities even in the presence of O_2, which are referred to as O_2-tolerance.

Herein, we focus on membrane-bound [NiFe]-hydrogenases (MBHs), which are O_2-tolerant hydrogenases. The 3D structures and active sites of MBHs are very similar to those of the standard hydrogenases except for the transition metal (iron) binding site that are located in the proximity of the catalytic center where Ni and Fe are bound, which are referred to as the proximal and [NiFe] active clusters, respectively (**Figure 2**).

Although the mechanisms of the O_2-tolerance still remained to be resolved, the structural differences of the proximal clusters between MBHs and the standard hydrogenases were suggested to be responsible for the O_2-tolerance mechanisms. More specifically, the proximal cluster of the standard hydrogenase contains [4Fe-4S]-4Cys cluster, while that of MBH contains [4Fe-3S]-6Cys cluster (**Figure 2**). In fact, two cysteine residues in MBH are replaced with glycine (Gly) residues in the standard [NiFe]-hydrogenases.

In addition, for the proximal cluster of MBH, three charge states have been reported; that is, the reduced, oxidized, and superoxidized states. Moreover, the 3D structure of the proximal cluster in MBH is also changed depending on those redox states. Moreover, combined crystallographic and spectroscopic analyses have recently suggested that a hydroxyl ion (OH^-) was attached to a Fe ion in the superoxidized states of the proximal cluster in *Ralstonia eutropha* MBH [30]. However, its functional role has still remained to be clarified.

In this section, we first introduce the way of how we investigated the electronic structures of the proximal cluster of *Ralstonia eutropha* MBH. Herein, systematic exploration of spin combinations of the proximal cluster was essential to obtain the reliable calculation data. For the calculation models, we constructed two structural models of the proximal cluster in the presence and absence of the hydroxyl ion and compared their detailed electronic structures, to reveal the functional role of the hydroxyl ion. To evaluate the total energy of various spin states and the electronic

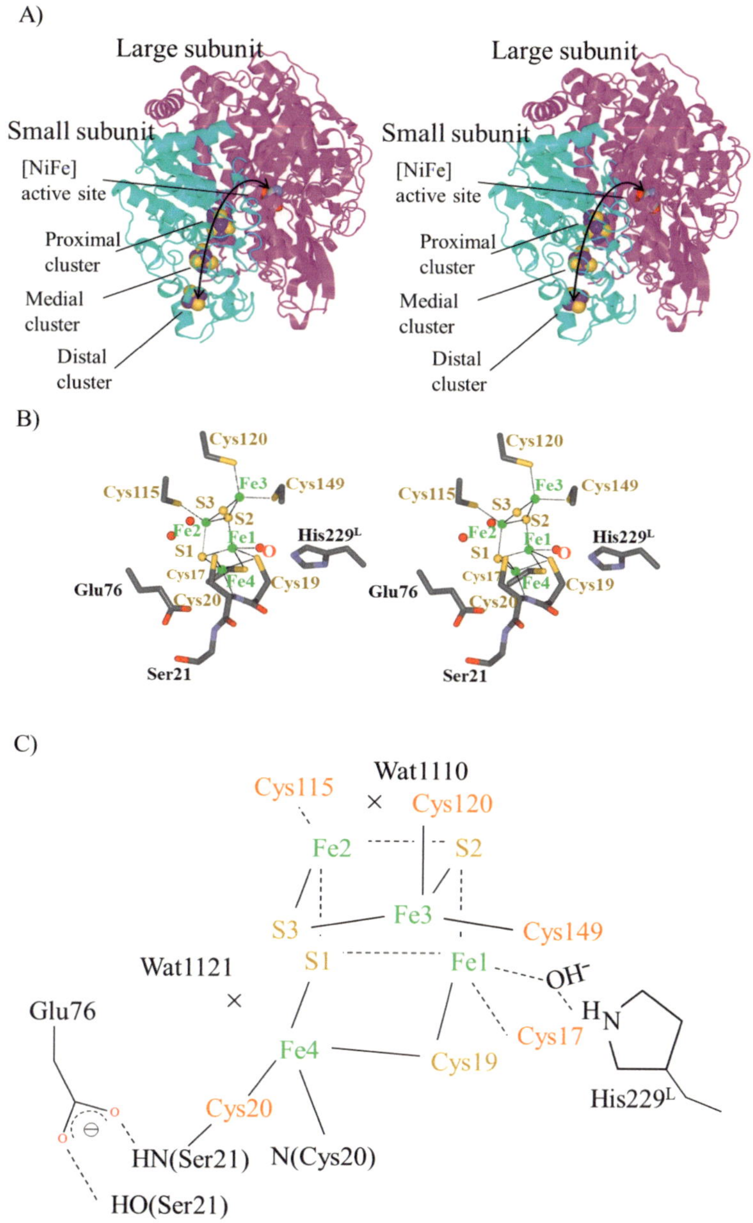

Figure 2.
Stereoview of the 3D structures of the entire structure (A) and proximal cluster (B) of Ralstonia eutropha *MBH, and schematic representation of model 2 (C). © Kim et al. [31].*

structures of the optimum energy states in the presence and absence of the hydroxyl ion, we employed full (*ab initio*) QM calculations with the use of the B3LYP functional as mentioned above (see Section 1.3).

2.2 Exploration of spin assignments

To build structural models, we employed the atomic coordinates of the proximal cluster in the superoxidized state of the crystal structure of *Ralstonia eutropha* MBH (PDB ID: 4IUD). Our computation models included the iron-sulfur cluster (i.e.,

[4Fe-3S]) and six Fe-coordinated cysteine (Cys) residues (i.e., Cys17, Cys19, Cys20, Cys120, Cys115, and Cys149). Moreover, we also included three amino acid residues (i.e., Ser21, Glu76, and His229), two crystal water molecules, and OH^- ion, all of which are coordinated to the iron-sulfur cluster.

Six amino acid residues (Cys17, Cys115, Cys120, Cys149, Glu76, and His229) were truncated by Cα atoms with the attachment of methyl groups ($\text{—}CH_3$). As mentioned in the last section, we constructed another similar model that did not include the OH^- ion to reveal its effects, and thus 103 and 101 atoms were included in our structural models. These are referred to as the original (model 2) and $\Delta(OH^-)$ (model 1) models, respectively (**Figure 2**).

Spectroscopic experiments elucidated the formal charge and total spin of the [4Fe-3S] cluster in the superoxidized state as +5 and 1/2, respectively [32]. For each of the iron and sulfur ions in the [4Fe-3S] proximal cluster, we set the formal charge as Fe^{2+} or Fe^{3+}, and S^{2-}, as found in the previous study [32]. Thus, the core consists of three Fe^{3+}, one Fe^{2+}, and $3S^{2-}$ ions.

Then, we constructed simple small fragmental models that were extracted from the 3D structure of the proximal cluster core: each of three Fe ions labeled as Fe2, Fe3, and Fe4 form a tetrahedral structure in the [4Fe-3S] proximal cluster, while the other Fe ion labeled as Fe1 form bipyramidal structures together with S_{Cys19}, S_{Cys17}, S1, (O atom of the hydroxyl ion), and S2 atoms (**Figure 2**).

Thus, we built small models including only the core atoms (i.e., Fe1, S_{Cys19}, S_{Cys17}, S1, and S2) in the presence and absence of the OH^- and evaluated the total energy values of the models. The analysis revealed that the optimum spin states with the minimum total energies were composed of the high spin states of Fe ions, which is consistent with the previous experimental data [33].

Herein, to specify the spin assignments of the [4Fe-3S] cluster, we represent them employing the nomenclature, BSij; that is, BS is an abbreviation of the broken symmetry state, and i and j indicate the (serial) numbers of Fe ions, as follows. Due to the two constraints, that is, (1) Fe ions take the high spin states as found above, and (2) the total spin sum is 1/2 (experimental data), the possible spin combinations of Fe ions are deduced as 5/2, −4/2, and − 5/2. We adopt the indices i and j that should be corresponding to the spin states of −4/2 and −5/2 (of Fe), respectively [34]. For example, BS12 represents that spin of Fe1 and Fe2 are assigned to the −4/2 and −5/2, respectively, and thus, 5/2 spin state is assigned to Fe3 and Fe4. Thus, BS12 represents (Fe1, Fe2, Fe3, Fe4) = (−4/2, −5/2, +5/2, +5/2).

Based on these considerations, we found that 12 spin assignments are possible for each structural model, and thus, we performed 24 QM calculations, to identify the optimal spin states of the [4Fe-3S] proximal cluster in the presence and absence of the hydroxyl ion. We employed the Gaussian09 package for all QM calculations with the B3LYP functional [3, 35]. For the [4Fe-3S] core and atoms that are coordinated to the Fe ions, the triple-ζ valence polarized (TZVP) basis set [36, 37] was adopted, and for the other atoms, the 6-311G** basis set [38] was employed. We performed geometry optimization with all hydrogen atoms being movable.

As a result of the analysis, we found that the total energy of BS12, BS21, BS13, and BS31 was smaller than the other states in model 1 and that the total energy of BS12, BS21, BS34, and BS43 was lower than the other states in model 2. Thus, we indicated that the favorable spin assignments were depending on the presence and absence of hydroxyl ion in the proximal cluster.

Note here that in previous studies employing DFT calculations and the simple iron-sulfur clusters, such as [2Fe2S], [3Fe4S], and [4Fe4S], BSij and BSji were shown to be identical [39, 40]. However, this equivalence of BSij and BSji was not satisfied in the present case, since the [4Fe-3S] proximal cluster in the MBH is distorted, compared with the simple Fe-S clusters that were analyzed in the previous studies. Moreover, the attachment of the hydroxyl ion (model 2) induced the distinct electronic structures when we compared with those of model 1 and the standard iron-sulfur clusters. Thus, due to the distorted geometrical structure and attachment of the hydroxyl ion, the equivalence of BSij and BSji cannot be assured in the present case. In fact, BS34 and BS43 of models 1 and 2 were definitely different in the total energy by 7.78 and 1.77 kcal/mol, respectively (here, the optimum total energy is set to 0 kcal/mol).

In this manner, we determined the optimal spin states of the proximal cluster of the MBH in the presence and absence of the hydroxyl ion, as the lowest energy states, that is, BS12 and BS34 of models 1 and 2, respectively. In the subsequent part, we describe the electronic structures of these spin states.

2.3 Functional role of OH^-

In the presence of O_2, the inactive form is induced with respect to the [NiFe] catalytic site of MBHs (i.e., the Ni-B state). For the reactivation of the catalysis, the [NiFe] active site is required to be changed to another state (i.e., the Ni-SI state) [41]. Two recovery mechanisms have been suggested up to date. First, Volbeda et al. [42] suggested that formation of a dimer enhanced the reactivation of MBHs; that is, one that is inactivated can be recovered by the other that is activated in the dimer. In this mechanism, at least two electrons should be transferred from the activated MBH to the inactivated MBH, and then the received electrons are transferred through the distal, medial, proximal clusters, and the [NiFe] active site [42].

In the second mechanism suggested by Kurkin et al. [43], the reduction of the [NiFe] active site in the presence of H_2 reactivates the inactive MBH, although the process requires a few seconds (note here that the O_2-sensitive hydrogenases such as the standard [NiFe] hydrogenase require over 1 h to be reactivated). In this reactivation process, H_2 cleavage reaction induces the Ni-SI state from the Ni-B state of the [NiFe] active site, and four electrons should be transferred from the [NiFe] active site to the proximal and medial clusters [43]. Notably, the directions of the electron transfers (ETs) are opposing between these two mechanisms that have been suggested so far.

To describe the differences of the electronic structures of the proximal cluster in the presence (model 2) and absence (model 1) of the hydroxyl ion [31], we focus on the lowest unoccupied molecular orbital (LUMO) here, since LUMO could be closely related to the ET, which is required to occur in both reactivation mechanisms as mentioned above. In fact, comparison of the LUMOs of models 1 and 2 led us to identify the effects of the hydroxyl ion on the LUMOs: In the absence of the hydroxyl ions (model 1), the LUMO was localized on Fe4 and S_{Cys20}, whereas in the presence of the hydroxyl ion (model 2), the LUMO was delocalized on S_{Cys17}, S_{Cys19}, S_{Cys20}, N_{Cys20}, the hydroxyl ion, Fe1, Fe2, and Fe4 (**Figure 3A** and **B**).

To further approach the substantial meanings of differences found in the calculated electronic structures, we also investigated the ET pathways by adopting an empirical simulation methodology, which can identify possible ET pathways by minimizing the penalties of the steps mediated by covalent bonds, hydrogen bonds, and spaces [44, 45]. Based on the aforementioned two ET processes, which have been suggested so far, that is, (1) the medial to proximal clusters and (2) the [NiFe] active site to proximal cluster, we examined the validity of these two ET pathways.

As a result of the analysis, we revealed that the optimal ET pathways were significantly overlapped with the delocalized LUMOs of model 2, which means that

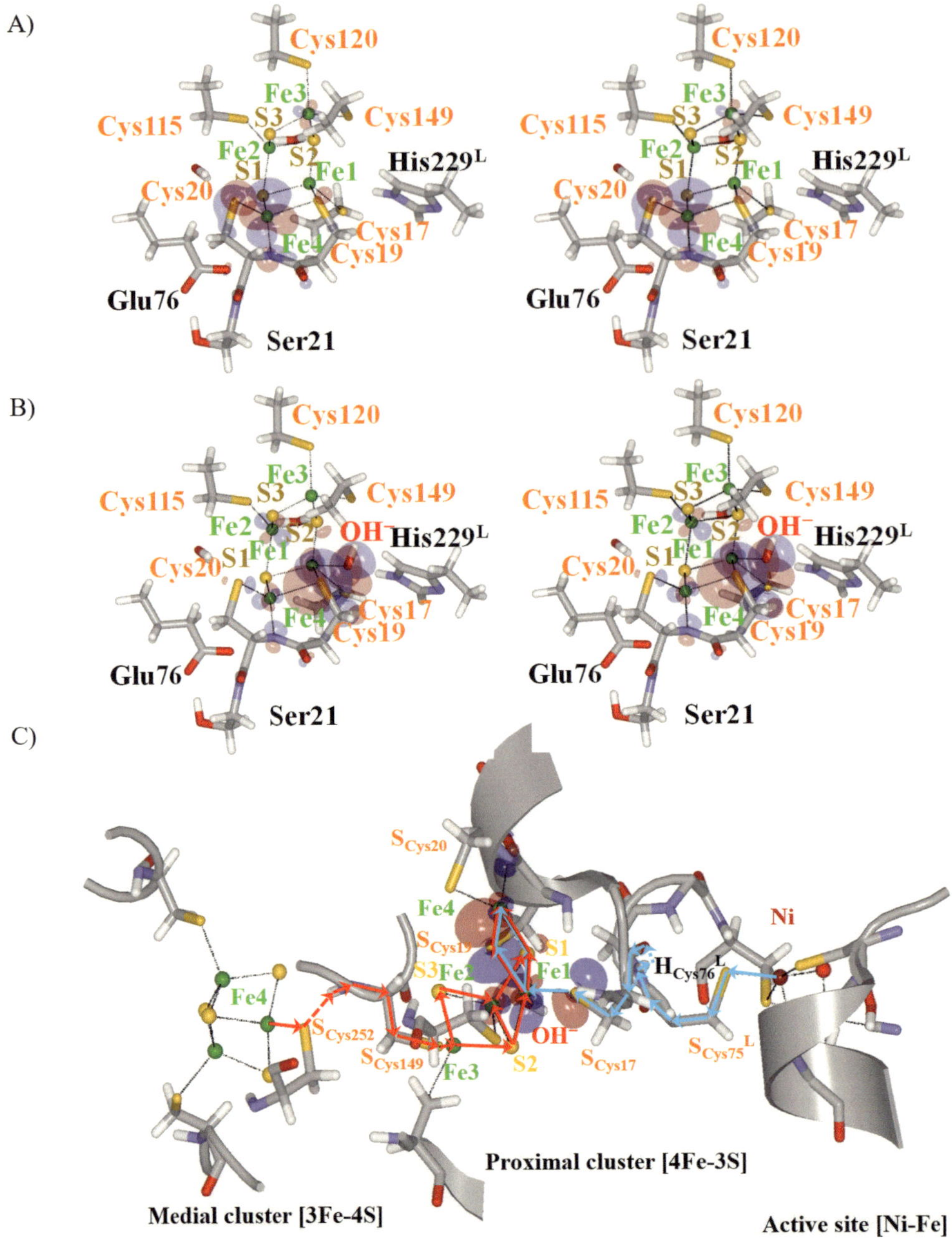

Figure 3.
Stereoview of the LUMOs in terms of the optimal spin states of models 1 (BS12) (A) and 2 (BS34) (B). In model 1 (i.e., $\Delta(OH^-)$), the LUMOs are localized on Fe4 and S_{Cys20} atoms. By contrast, in model 2 (i.e., involving OH^-), the LUMO is distributed and thus principally composed of S_{Cys17}, S_{Cys19}, S_{Cys20}, N_{Cys20}, the hydroxyl ion, Fe1, Fe2, and Fe4. To investigate the relationships between the effects of the hydroxyl ion and the ET mechanisms, the plausible ET pathways obtained by an empirical method (i.e., pathway*) to search for the ET pathways (C) are compared with the distribution of the LUMO of model 2, which is identical to that shown in panel (B) (note that the directions for rendering of the structures are identical in the panels (A) and (B), while those are different between the panels (A)/(B) and (C)). The blue line with an arrow shows the ET pathway from the [NiFe] active site to the proximal cluster, and the red line with an arrow shows the ET pathway from the medial to proximal clusters. © Kim et al., [31].*

the attachment of the hydroxyl ion to Fe1 may promote the ETs (**Figure 3C**). This also means that the attachment of the hydroxyl ion creates the ET pathways in the proximal cluster by inducing the electron delocalization of the LUMO, thus forming the S_{Cys17}-HO^--S_{Cys19}-Fe4 segmental electronic field, which is a main component of the delocalized LUMO in the presence of the OH^-.

The origin and formation mechanisms of the delocalized LUMO are a very interesting issue. However, the limitation of space here does not allow us to describe it, and so for the detailed descriptions on this issue, see text and **Figure 6(a)** in our report [31].

To the best of our knowledge, this is the first work to have revealed the mechanisms of creation of the ET pathways in biological macromolecular systems [31]. Note here that a tiny molecular species, that is, OH^-, is a trigger to generate the ET pathways. Thus, organisms regulate the functions employing such a subtle factor but thereby dramatically change their physiological status. The present achievements could further be a solid basis toward sophisticated rational design of novel catalysts, reactions, and functional materials, through regulations of the elaborately constituted orbital-based electronic structures.

3. Hybrid *ab initio* QM/MM molecular dynamics simulation

In the last section, we presented the effects of the attachment of a hydroxyl ligand on the electronic structure of the proximal cluster in the MBH and also showed that the effects could be closely related to the ET and recovery from the inactive form of the [NiFe] active site (i.e., the O_2-tolerance of the MBH). Thus, the tiny species dramatically changes the electronic structures of the enzymes, as discussed above.

In this section, we demonstrate that electronic structures are dynamically changed as catalytic reactions proceed. In order to investigate such dynamical properties of electronic structures in biological macromolecular systems, we built a hybrid *ab initio* QM/MM molecular dynamics (MD) calculation system on superparallel computers. This computational system enabled us to evaluate dynamical transitions of electronic structures in catalytic sites involving the environmental structures such as protein moieties, nucleic acid (RNA and DNA) moieties, membrane moieties, and solvent [17, 22, 46, 47].

We introduce editing reaction of aminoacyl-tRNA synthetase (aaRS), which forms a protein family composed of twenty enzymes, divided into two classes, that is, classes I and II [48]. Here, we focus on two class I aaRSs, that is, leucyl-tRNA synthetase (LeuRS) and valyl-tRNA synthetase (ValRS). We performed hybrid *ab initio* QM/MM MD simulations and thereby revealed that as the reactions proceeded, dynamical rearrangements of molecular orbitals (MOs) occurred, which was critical for both covalent bond formation and cleavage. We emphasize here that such dramatic transitions of electronic structures would be characteristic in catalytic reactions of biological macromolecular systems, as also indicated in hydrogenase.

3.1 Aminoacyl-tRNA synthetases (aaRSs)

In the central dogma, which demonstrates the flow of the genetic information (e.g., from gene coded in genome DNA toward protein), the adaptation between a codon (i.e., combination of three nucleotide bases) and an amino acid (aa) is required in the protein biosynthesis (i.e., translation), while synthesis of messenger RNA (mRNA) (i.e., transcription) occurs based on the rules of base pairing (e.g. the Watson-Crick and wobble base pairs). Thereby, base sequences of genes are converted to amino acid sequences of proteins. Here, aaRSs play a critical role by correctly recognizing and attaching both specific amino acid and cognate tRNA, thereby forming aminoacyl-tRNA (i.e., aa-tRNA^{aa}). This crucial catalytic reaction is referred to as aminoacylation.

If aaRSs misattach noncognate amino acids (or tRNAs) to the specific tRNAs (or amino acids), then the noncognate amino acids are incorporated into the protein as an incorrect amino acid residue, which is different from the correct one coded in the gene, since the tRNA recognizes its cognate codon by making base pairs between the codon (in mRNA) and the anticodon moiety of the tRNA. It should be noted here that only aaRSs determine the relationship between amino acids and codons for biosynthesis of proteins.

Based on the 3D structures, aaRSs are classified into two classes, that is, classes I and II [49, 50]. More specifically, catalytic cores of class I aaRSs contain the classical nucleotide-binding fold (i.e., the Rossmann fold). By contrast, the active sites of class II aaRSs possess an antiparallel β sheet flanked by α helices, which is the architecture completely different from the Rossmann fold. Based on the structural similarity of the catalytic and noncatalytic domains, each class is further classified into three subclasses a, b, and c [51, 52].

3.2 Editing reaction of aaRSs

As mentioned above, high translational fidelity is essential in the decoding of genetic information from mRNA (i.e., base sequences) to protein (i.e., amino acid sequences). Notably, the aminoacylation reaction of aaRSs consists of two steps; that is, (1) activation of the amino acid yielding aminoacyl adenylate and (2) transfer of amino acid moiety of the aminoacyl adenylate to the 3′-end of the tRNA. However, misactivated amino acids and misaminoacylated tRNAs are generated occasionally, since some amino acids are structurally similar (e.g., leucine (Leu), isoleucine (Ile), and valine (Val)).

To ensure the translational fidelity, some aaRSs possess editing functions to correct such errors [53–60]. Correspondingly, two types of editing reactions are known, that is, pre- and posttransfer editing reactions, which hydrolyze a misactivated amino acid and misaminoacylated tRNA, respectively. Herein, we focus on the posttransfer editing reaction, since the catalytic mechanisms were remained to be elucidated for the last some decades.

3.3 *Ab initio* QM/MM MD simulation of editing reaction

For the Leu system, we constructed a structural model of LeuRS in complex with a misaminoacylated tRNA^{Leu} (i.e., valyl-tRNA^{Leu}), where the 3′-end nucleotide (adenine 76; A76) is bound to the active site for the editing reaction in the connective polypeptide (CP) 1 domain [22] (**Figure 4**). Then, we performed classical MD simulation and succeeded in identification of the nucleophilic water for the editing reaction [23]. Employing this structural model, we performed hybrid *ab initio* QM/MM MD simulations with the use of our QM/MM interface program [47] that connects QM and MM calculation engines (i.e., GAMESS [61] and AMBER [62], respectively).

To determine the reaction path, we employed an adiabatic mapping approach, in which hybrid *ab initio* QM/MM MD simulations were performed to enhance the conformational sampling, together with hybrid *ab initio* QM/MM geometry optimization being employed to reach the potential energy surface. This scheme enabled us to conduct more effective explorations of both conformations and electronic structures than previous schemes that employ only geometry optimizations [63]. We assumed some possible reaction pathways and performed hybrid *ab initio* QM/MM MD simulation for each pathway, which provided the estimation of the energy barrier

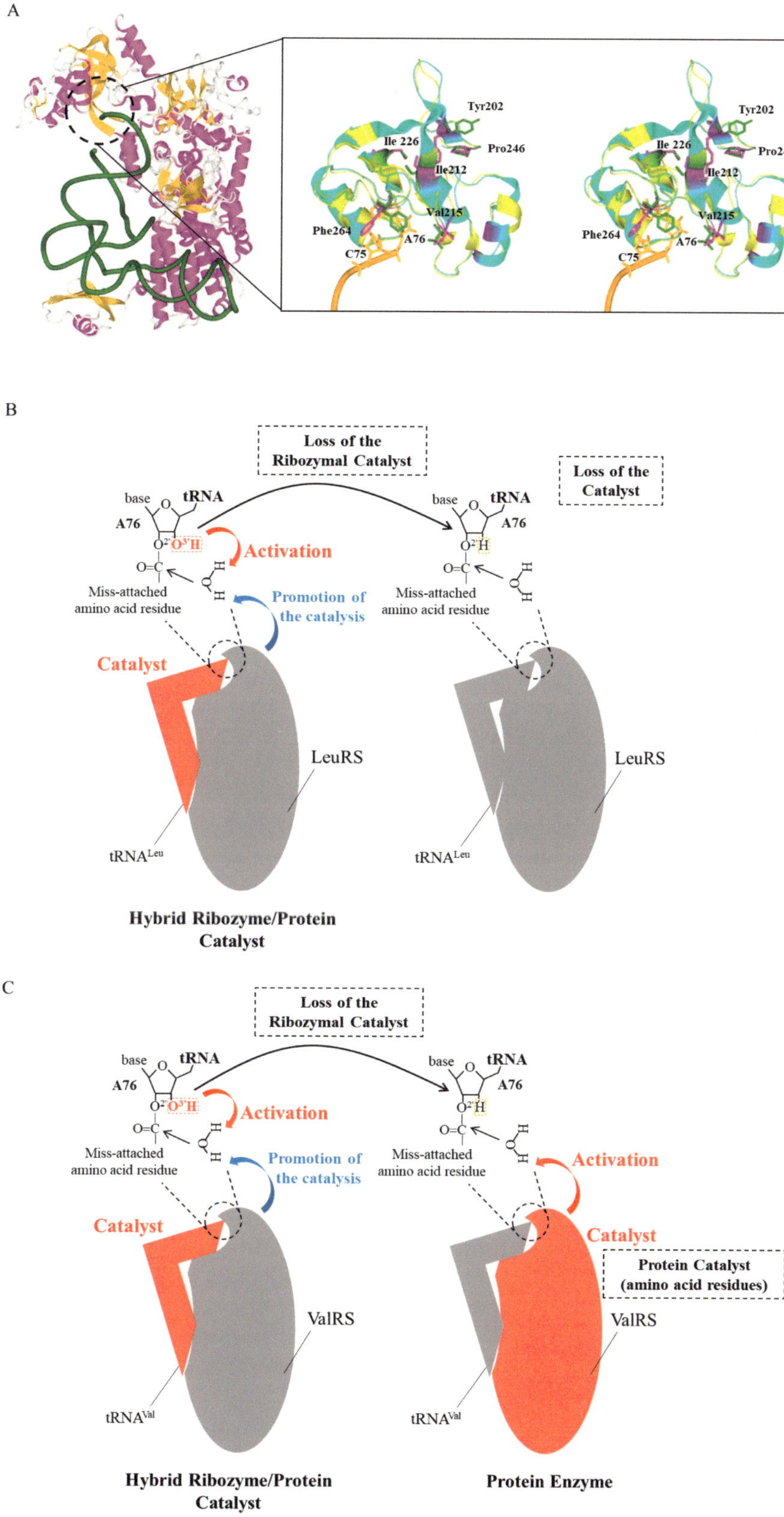

Figure 4.
(A) (Left) Entire 3D structure of the ValRS·thereonyl-tRNA^{Val} complex is shown as an example of structures of class Ia aaRSs (i.e., involving the Leu, Val, and Ile systems). (Right) Catalytic site of the editing reaction is

shown (stereoview). The crystal structure of the complex (1IVS) is colored yellow (for amino acid and RNA backbones), green (for amino acid side chains), and orange (for nucleic acids). The crystal structure of the isolated CP1 domain (1WK9) is colored light blue (for amino acid backbone) and magenta (for amino acid side chains). (B, C) Schematic representations of fundamental reaction schemes of hybrid ribozyme/protein catalyst (left) and their variant systems. The black circles (broken line) show the catalytic site, and the macromolecules involving the catalysts are colored red. The mechanisms of the editing reactions in both Leu and Val systems are revealed to be common by our recent studies: Interestingly, the editing reaction is ribozymal together with assists of protein moiety (left panels in (B-C)), which is referred to as hybrid ribozyme/protein catalysis [63]. The ribozymal factor (i.e., 3′-OH of A76 of tRNA) activates the nucleophilic water molecule (as represented by the red arrow), and the protein (LeuRS and ValRS) moiety promotes the catalysis (the blue arrow) [63]. Nevertheless, in the "defective" mutated systems (i.e., replacements of the aforementioned 3′-OH with 3′-H), reductions of the editing activities were experimentally revealed to be distinct. In the Leu system (B), the editing activity significantly decreased by the mutation, whereas in the Val and Ile systems, those were preserved. Our modeling studies elucidated that these were due to the absence and presence of compensation mechanisms in the Leu and Val/Ile protein moieties, respectively. Actually, in our previous study, we constructed the atomistic structural model of the Val system (i.e., the ValRS·threonyl-tRNA^{Val} (misaminoacylated) complex) and showed that for the Val system with a "defective" ribozymal activator of tRNA^{Val}, the protein (ValRS) moiety could activate the nucleophilic water molecule (the red arrow), which is referred to as the protein enzyme (note that the definition of protein enzyme described here is different from that employed by Cech [64]). As discussed in the text, this transition from the hybrid ribozyme/protein catalyst toward the protein enzyme may fill a gap found in the evolutionary transition from the RNA world to the current RNP world, which could possibly occur in primordial biological macromolecular systems. © Sakabe et al., [25].

depending on the reaction pathway. Thus, we determined the optimal reaction pathway and thereby elucidated the mechanism of the editing reaction (**Figure 4**).

As a result of the analysis, we discovered a novel catalytic mechanism, as follows: the editing reaction was revealed to be driven by the $O^{3'}$ of the ribose moiety of the 3′-end nucleotide A76, which acts as the general base to activate the nucleophilic water. Surprisingly, the editing of the LeuRS·valyl-tRNA^{Leu} complex was revealed to be ribozymal [63, 65, 66].

Furthermore, we found that this ribozyme reaction was enhanced by protein, through the formation of a hydrogen bond with the catalytic core of tRNA^{Leu}. Since the catalytic cores of the conventional protein-dependent ribozymes such as ribosome and group I intron [67] are purely composed of RNAs [64], this finding, that is, direct contributions of the protein moiety on the ribozymal reaction, is novel, and thus, we referred to the LeuRS·valyl-tRNA^{Leu} complex as a "hybrid ribozyme/protein catalyst" (**Figure 4**).

A very recent experimental study conducted by Dulic et al. experimentally showed that the defective mutation of the $O^{3'}$ atom (i.e., 3′-OH of A76 was replaced with 3′-H) significantly reduced the activity of the Leu system ($\sim 10^4$-fold rate reduction) (**Table 1**) [68], and thus, the hybrid ribozyme/protein catalyst mechanism has been reasserted by the biochemical experiments.

Can we generalize this novel, hybrid ribozyme/protein catalytic reaction mechanism of the LeuRS·valyl-tRNA^{Leu} complex? Considering structural similarity of class Ia aaRSs, which involves LeuRS, valyl-tRNA synthetase (ValRS), and isoleucyl-tRNA synthetase (IleRS), they may share a common editing mechanism with LeuRS. However, an experimental conflict has still been left to be resolved as follows. While the aforementioned modification of the O3′ reduced the editing reaction in the Leu system [63, 68], the identical modification was not severe in the reduction of the editing activity with respect to the Val and Ile systems (**Table 1**) [69].

To resolve this discrepancy of the experiments, we constructed a structural model of the complex of ValRS and misaminoacylated tRNA [63] and thereby suggested that the hybrid ribozyme/protein catalyst mechanism was also shared in the editing reaction of the Val system [25]. Furthermore, to explain how the variant Val system can maintain its catalytic activity (**Table 1**), we constructed a structural model involving the variant tRNA^{Val} in which the 3′-OH (reactive) of A76 was

	Class	Species	Attached site	Activator	Reduction rate	Reference
LeuRS	Ia	*T. thermophirus*	$O^{2\prime}$	3′-OH	—	
LeuRS	Ia	*E. coli*	$O^{2\prime}$	Δ(3′-OH)[a]	10^4-fold	[68]
LeuRS (D342A)	Ia	*E. coli*	$O^{2\prime}$	3′-OH	3-fold	[68]
ValRS	Ia	*E. coli*	$O^{2\prime}$	Δ(3′-OH)[a] and D276	10-fold	[69]
IleRS	Ia	*E. coli*	$O^{3\prime}$	Δ(2′-OH)[b] and E327	5-fold	[69, 70]
ProRS	IIa	*E. coli*	$O^{3\prime}$	Δ(2′-OH)[b]	10^3-fold	[71]
ThrRS	IIa	*E. coli*	$O^{3\prime}$	Δ(His73)[c,d]	10^4-fold	[72, 73]
ThrRS	IIa	*Pyrococcus abyssi*	$O^{3\prime}$	2′-OH	—	[74, 75]
PheRS	IIc	*E. coli*	$O^{2\prime}$	Δ(3′-OH)[a]	300-fold	[76]

© Sakabe et al., [25].
[a]*Δ(3′-OH) represents the replacement of 3′-OH group of A76 with 3′-H atom.*
[b]*Δ(2′-OH) represents the replacement of 2′-OH group of A76 with 2′-H atom.*
[c]E. coli *ThrRS acts as a protein enzyme.*
[d]Escherichia coli *ThrRS acts as a protein enzyme (see* ***Figure 4C****).*

Table 1.
Summary of the nucleophile activators and the reduction rate of hybrid ribozyme/protein catalysts (i.e., LeuRS, ValRS, IleRS, PheRS, ProRS, Pyrococcus abyssi *ThrRS, and ribosome) and a variant aaRS (D342 of* E. coli *LeuRS is replaced with Ala).*

replaced with the 3′-H (unreactive). Based on this analysis [25], we suggested that the editing activity of the variant Val system is maintained by an amino acid residue of ValRS acting as the possible general base (**Figure 4**). More specifically, Asp276 of ValRS can replace the functional role of 3′-OH of A76 (i.e., activation of the nucleophilic water).

In this manner, the hybrid ribozyme/protein catalysts are operated in both Leu and Val systems. Moreover, this suggested that in the "defective" Val system that contains "unreactive" ribozymal functional group (i.e., 3′-OH of A76 is replaced with 3′-H), the function of a hybrid ribozyme/protein catalyst can be transferred to a protein enzyme. This could also be related to the evolutional transition from RNA enzymes (the RNA world) to protein enzymes (assisted by RNA) (the RNP world), intermediated by hybrid ribozyme/protein catalysts.

We further suggested that the ribozymal mechanism that we discovered is common in the editing reaction of various aaRS systems beyond the classes (**Table 1**) [63]. In fact, *Thermus thermophilus* IleRS (class Ia) [77], *Pyrococcus abyssi* ThrRS (class IIa) [72, 74, 75, 78], and *Enterococcus faecalis* ProRS (class IIa) [71] showed the similar binding mode of the nucleophilic water in the catalytic site. Furthermore, Kumar et al. also suggested that the editing reaction of the complex of prolyl-tRNA synthetase (ProRS) and alanyl-$tRNA^{Pro}$ exhibited a similar mechanism, in which the 2′-OH group of A76 of $tRNA^{Pro}$ was involved in the substrate binding and the activation of the nucleophilic water [71].

These data are summarized in **Figure 4** and **Table 1**. In this section, we further discuss the dynamical aspects of the electronic structures in the editing reaction of Leu and Val systems, investigated by hybrid *ab initio* QM/MM MD simulations of the LeuRS·valyl-$tRNA^{Leu}$ and ValRS·threonyl-$tRNA^{Val}$ complexes, respectively.

3.4 Dynamic rearrangement of MOs in the editing reactions

For both Leu and Val systems, we suggested that the editing reactions occur in a similar manner [25, 63]. Actually, in both systems, the reactions were shown to be initiated by opening of the "H-gate": The H-gate is defined by a dihedral angle, $C^{4\prime}$–$C^{3\prime}$–$O^{3\prime}$–$HO^{3\prime}$ of A76, and its opening represents the rotation of the dihedral angle by ~100°, which thus leads to the nucleophilic attack of the water molecule.

For the Val system, employing the hybrid *ab initio* QM/MM calculations, we evaluated the electronic structures for the two distinct H-gate states, that is, the opened and closed states [25, 63]. The resultant data showed that the LUMO was located in the C atom in both closed and opened H-gate states (**Figure 5**). By contrast, the energy levels of the MOs that include the 2*p* orbital of oxygen atom of the nucleophilic water (O_W) (i.e., the "reactive" MO) were different depending on the two distinct states of the H-gate: In the opened H-gate state, we observed the MO as HOMO−6, while in the closed H-gate state, the MO was observed as HOMO−11, for which the energy level was much lower compared with that of the former state.

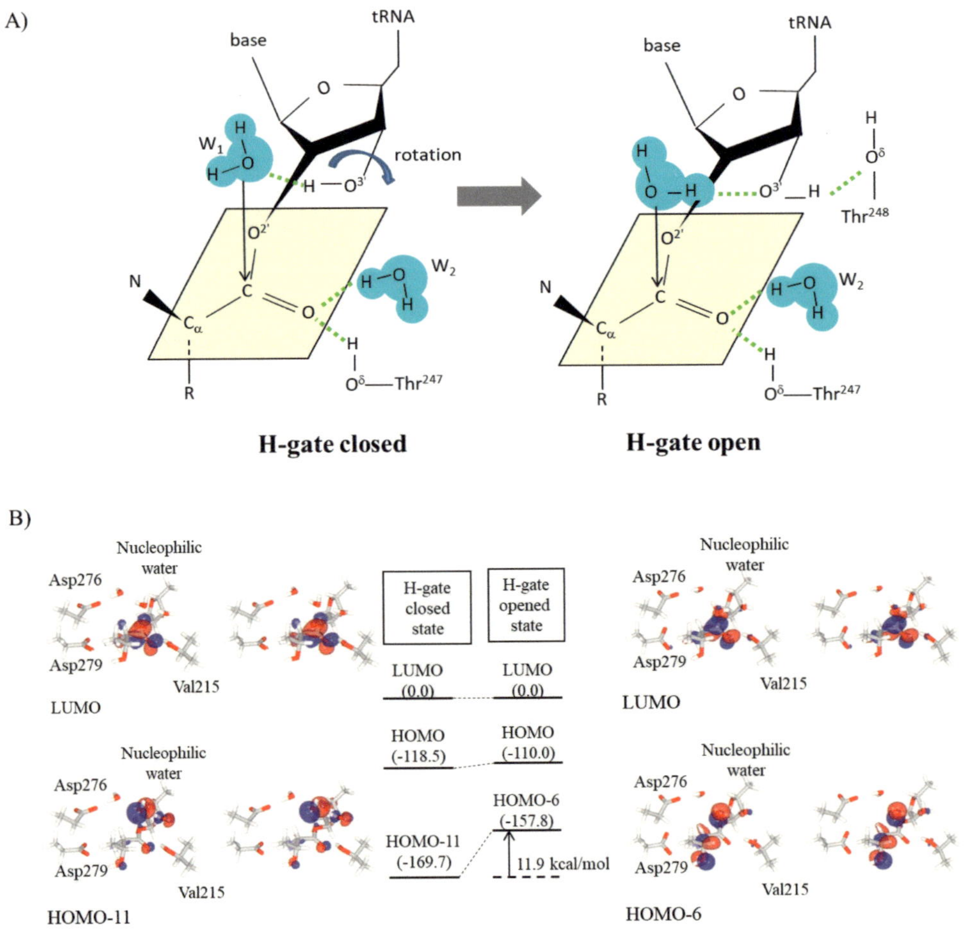

Figure 5.
Schematic picture of the H-gate closed and H-gate open states (A). Stereoview of LUMO and molecular orbitals (MOs) of the catalytic site including the nucleophilic water molecule for H-gate closed and H-gate opened conformations. HOMO-11 for H-gate closed and HOMO-6 for H-gate opened conformations. The ribose and threonine moieties of the substrate, Val215, Asp276, Asp279, and the three water molecules are included as the QM region. (B) Energy diagrams concerning the LUMOs and HOMO-11 (HOMO-6) in H-gate closed state (H-gate opened state). © Sakabe et al. [25].

In this manner, as the reaction proceeds, the energy level of the reactive MO seems to go up (**Figure 5**). Thus, if the nucleophilic attack would be achieved, the energy level of the reactive MO in the nucleophilic water could be raised up to that of the HOMO, which would thus result in hybridization of the HOMO and LUMO. In fact, the similar dynamical rearrangements in the electronic structure were also observed in our hybrid *ab initio* QM/MM calculations of the editing reaction occurring in the LeuRS·valyl-tRNA^{Leu} complex [65, 66].

For the Leu system, we further investigated the overall mechanism of the editing reaction employing the hybrid *ab initio* QM/MM MD simulations. Based on the orbital analysis of the trajectory of the hybrid MD simulations, we found more detailed dynamical properties of the electronic structures (**Figure 6**): In the initial stage of the editing reaction, the HOMO did not contain the 2*p* orbital of the O_w atom of the nucleophilic water, even though it attacked the C atom, which seemed to be inconsistent with the frontier orbital theory (FOT) [79].

However, when H-gate was open (state 3), the nucleophilic water approached the C atom, and the energy level of an MO that most contained the 2*p* orbital of O_w was elevated to HOMO−9 from the HOMO−14 observed in state 1. This elevation decreased the energy difference between the LUMO, which contained the reactive moiety (i.e., atomic orbitals of the carbonyl group of the substrate and the $O^{2\prime}$ atom

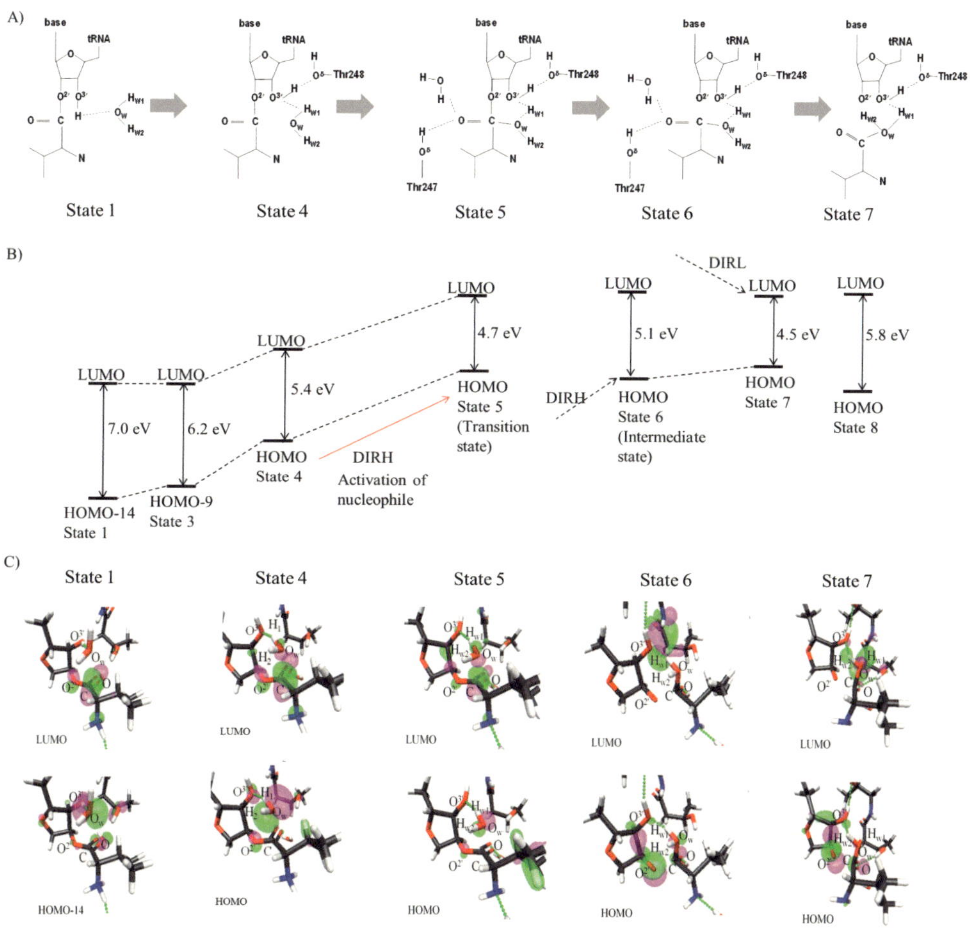

Figure 6.
Schematic representation of the reaction states of editing mechanism of LeuRS system is shown in (A). Energy diagram of the editing (B) and molecular orbitals (C) of states 1, 4, 5, 6, and 7 are shown. © Kang et al., [66].

of A76), and the MO involving the $2p$ orbital of O_w, from 7.0 eV to 6.2 eV. In state 4, this energy gap further decreased to 5.4 eV, and the MO involving $2p$ orbital of O_w became the HOMO, while the LUMO remained to be localized on the C atom.

In this stage, both HOMO and LUMO were the reactive MOs responsible for the catalytic reaction, and thus, the electronic structure was fully consistent with the FOT. In fact, the Ow−C covalent bond was formed by the interaction between the $2p$ orbital of O_w (HOMO) and the $2p$ orbital of C (LUMO). Here, we referred to such dynamical rearrangements of the electronic structure as the dynamical induction of the reactive MOs for the HOMO and LUMO, that is, DIRH and DIRL, respectively [65, 66].

In our preliminary studies of other biological macromolecular systems, we have also observed the DIRH and DIRL mechanisms (unpublished data). Future theoretical and experimental analyses are amenable to examine the generality of this picture on dynamical rearrangements of electronic structures occurring in the reaction cycles of biological macromolecular systems. In addition, aaRSs are closely relevant to the molecular evolution and the origin of life, which should thus be considered from *ab initio* QM calculations. So, the present achievements are also related to those evolutional issues, although we do not describe them herein due to space limitation (for more details, see the literature [25]).

4. Conclusions

In this chapter, we introduced investigations employing *ab initio* QM calculations and hybrid *ab initio* QM/MM MD calculations. For the latter, a catalytic reaction site is considered at *ab initio* QM level, and the other parts, such as the remainder in protein structures and solvent water molecules, are considered at MM (i.e., classical) level, and thus, we can consider the entire system with reasonable computational costs, to evaluate the electronic structure of the catalytic active site. In both analyses of biological macromolecular systems, we revealed the significant reconstitutions of the electronic structures in the reaction cycles.

In the first example, we demonstrated the detailed electronic structures of a crucial functional site, the proximal cluster, in the MBH, which contains multiple transition metals as [4Fe-3S] and is closely related to the ET. We analyzed the effects of the OH^- ion that was experimentally identified in the proximal cluster, to the ET mechanisms. Thereby, we revealed that the OH^- ion created the ET pathways by inducing the delocalization of the LUMO of the proximal cluster.

This means that tiny molecular species (e.g., OH^-) can induce dramatic rearrangements of the electronic structure in the biological macromolecular systems, which thus generates the ET pathways. This is the first work to point out the mechanisms to create the ET pathways in biological macromolecular systems. In this manner, organisms regulate the biological functions employing such a subtle factor but thereby dramatically change their physiological status.

In the second example, we investigated dynamical changes of the electronic structures in the catalytic reaction cycles of the LeuRS·valyl-$tRNA^{Leu}$ and ValRS·-threonyl-$tRNA^{Val}$ complexes, employing our hybrid *ab initio* QM/MM MD calculation system, which is a state-of-the-art theoretical technique to elucidate the functional mechanisms of biological macromolecular systems. As a consequence, we revealed that the dynamical geometrical changes induced the dramatic rearrangemets of the electronic structures.

Thereby, the reactive MOs, which are positioned energetically far from the Fermi levels in the initial stages of the reaction cycles, are dynamically rearranged, but those MOs become the HOMO and LUMO, as the reaction cycles proceed. Thus, this reaction stage is followed by the subsequent phase of the covalent bond formation and cleavage. The obtained picture could be found in functional mechanisms of other various biological macromolecular systems, and thus, the generality of the presented novel picture is further amenable to future theoretical and experimental works. Thereby, this picture could be considered to be a characteristic feature in biological macromolecular systems.

Acknowledgements

This work was partly supported by Grants-in-Aid from the Ministry of Education, Culture, Sports, Science and Technology (MEXT), under contract Nos. 21340108 and 25287099. Computations were performed using computer facilities of the Computer Center for Agriculture, Forestry, and Fisheries Research, MAFF, Japan, and the Supercomputer Center, Institute for Solid State Physics, The University of Tokyo. JK was supported by the Korea Research Fellowship Program through the National Research Foundation of Korea (NRF) funded by the Ministry of Science and ICT (NRF-2017H1D3A1A01053094).

Conflict of interest

The authors declare to have no conflicting interests.

Author details

Jiyoung Kang[1], Takuya Sumi[2] and Masaru Tateno[2]*

1 Center for Systems and Translational Brain Sciences, Institute of Human Complexity and Systems Science, System Science Center for Brain and Cognition, Yonsei University, Seoul, Republic of Korea

2 Graduate School of Life Science, University of Hyogo, Kamigori-cho, Hyogo, Japan

*Address all correspondence to: tateno1611@gmail.com

References

[1] Becke AD. Density-functional thermochemistry .3. The role of exact exchange. The Journal of Chemical Physics. 1993;**98**(7):5648-5652

[2] Perdew JP, Ernzerhof M, Burke K. Rationale for mixing exact exchange with density functional approximations. The Journal of Chemical Physics. 1996; **105**(22):9982-9985

[3] Becke AD. Density-functional exchange-energy approximation with correct asymptotic-behavior. Physical Review A. 1988;**38**(6):3098-3100

[4] Vosko SH, Wilk L, Nusair M. Accurate spin-dependent Electron liquid correlation energies for local spin-density calculations - a critical analysis. Canadian Journal of Physics. 1980; **58**(8):1200-1211

[5] Yanai T, Tew DP, Handy NC. A new hybrid exchange–correlation functional using the coulomb-attenuating method (CAM-B3LYP). Chemical Physics Letters. 2004;**393**(1):51-57

[6] Iikura H, Tsuneda T, Yanai T, Hirao K. A long-range correction scheme for generalized-gradient-approximation exchange functionals. The Journal of Chemical Physics. 2001;**115**(8): 3540-3544

[7] Warshel A, Levitt M. Theoretical studies of enzymic reactions: Dielectric, electrostatic and steric stabilization of the carbonium ion in the reaction of lysozyme. Journal of Molecular Biology. 1976;**103**(2):227-249

[8] Bruice TC, Kahn K. Computational enzymology. Current Opinion in Chemical Biology. 2000;**4**(5): 540-544

[9] Field MJ. Simulating enzyme reactions: Challenges and perspectives. Journal of Computational Chemistry. 2002;**23**(1):48-58

[10] Field MJ, Bash PA, Karplus M. A combined quantum-mechanical and molecular mechanical potential for molecular-dynamics simulations. Journal of Computational Chemistry. 1990;**11**(6):700-733

[11] Lin H, Truhlar DG. QM/MM: What have we learned, where are we, and where do we go from here? Theoretical Chemistry Accounts. 2007;**117**(2): 185-199

[12] Mulholland AJ. Modelling enzyme reaction mechanisms, specificity and catalysis. Drug Discovery Today. 2005; **10**(20):1393-1402

[13] Ryde U. Combined quantum and molecular mechanics calculations on metalloproteins. Current Opinion in Chemical Biology. 2003;**7**(1):136-142

[14] Senn HM, Thiel W. QM/MM studies of enzymes. Current Opinion in Chemical Biology. 2007;**11**(2):182-187

[15] Senn HM, Thiel W. QM/MM methods for biomolecular systems. Angewandte Chemie, International Edition. 2009;**48**(7):1198-1229

[16] Gao J, Ma S, Major DT, Nam K, Pu J, Truhlar DG. Mechanisms and free energies of enzymatic reactions. Chemical Reviews. 2006;**106**(8): 3188-3209

[17] Kang J, Hagiwara Y, Tateno M. Biological applications of hybrid quantum mechanics/molecular mechanics calculation. Journal of Biomedicine & Biotechnology. 2012; **2012**:11

[18] Cavasotto CN, Adler NS, Aucar MG. Quantum chemical approaches in

structure-based virtual screening and Lead optimization. Frontiers in Chemistry. 2018;**6**(188)

[19] Kulik HJ. Large-scale QM/MM free energy simulations of enzyme catalysis reveal the influence of charge transfer. Physical Chemistry Chemical Physics. 2018;**20**(31):20650-20660

[20] Olsson MA, Ryde U. Comparison of QM/MM methods to obtain ligand-binding free energies. Journal of Chemical Theory and Computation. 2017;**13**(5):2245-2253

[21] Kang J, Tateno M. In: Pahlavani MR, editor. Recent Applications of Hybrid Ab Initio Quantum Mechanics – Molecular Mechanics Simulations to Biological Macromolecules. 2012

[22] Kang J, Ohta T, Hagiwara Y, Nishikawa K, Yamamoto T, Nagao H, et al. Electronic and geometric structures of the blue copper site of azurin investigated by QM/MM hybrid calculations. Journal of Physics. Condensed Matter. 2009;**21**(6):064235

[23] Hagiwara Y, Nureki O, Tateno M. Identification of the nucleophilic factors and the productive complex for the editing reaction by leucyl-tRNA synthetase. FEBS Letters. 2009;**583**(12):1901-1908

[24] Hagiwara Y, Nureki O, Tateno M. Structural modelling of the complex of leucyl-tRNA synthetase and mis-aminoacylated $tRNA^{Leu}$. FEBS Letters. 2009;**583**(4):825-830

[25] Sakabe K, Kang J, Terada R-i, Nishigami H, Tateno M. Possible transition from hybrid ribozyme/protein catalyst toward protein enzyme to compensate for defective catalytic activator. Journal of the Physical Society of Japan. 2018;**87**:124801

[26] Warshel A, Kato M, Pisliakov AV. Polarizable force fields: History, test cases, and prospects. Journal of Chemical Theory and Computation. 2007;**3**(6):2034-2045

[27] Jorgensen WL. Special issue on polarization. Journal of Chemical Theory and Computation. 2007;**3**(6):1877

[28] Vignais PM, Billoud B. Occurrence, classification, and biological function of hydrogenases: An overview. Chemical Reviews. 2007;**107**(10):4206-4272

[29] Friedrich B, Fritsch J, Lenz O. Oxygen-tolerant hydrogenases in hydrogen-based technologies. Current Opinion in Biotechnology. 2011;**22**(3):358-364

[30] Frielingsdorf S, Fritsch J, Schmidt A, Hammer M, Lowenstein J, Siebert E, et al. Reversible [4Fe-3S] cluster morphing in an O_2-tolerant [NiFe] hydrogenase. Nature Chemical Biology. 2014;**10**(5):378-385

[31] Kim J, Kang J, Nishigami H, Kino H, Tateno M. Ab initio electronic structure calculation of [4Fe-3S] cluster of hydrogenase as dihydrogen dissociation/production catalyst. Journal of the Physical Society of Japan. 2018;**87**(3):034804

[32] Pandelia ME, Bykov D, Izsak R, Infossi P, Giudici-Orticoni MT, Bill E, et al. Electronic structure of the unique [4Fe-3S] cluster in O_2-tolerant hydrogenases characterized by ^{57}Fe Mossbauer and EPR spectroscopy. Proceedings of the National Academy of Sciences of the United States of America. 2013;**110**(2):483-488

[33] Grajczyk R, Sean Muir RB, Sleight AW, Subramanian MA. Structural and magnetic investigation of Fe^{3+} and Mg^{2+} substitution into the trigonal bipyramidal site of $InGaCuO_4$. Journal of Solid State Chemistry. 2013;**199**:102-108

[34] Volbeda A, Amara P, Darnault C, Mouesca JM, Parkin A, Roessler MM, et al. X-ray crystallographic and computational studies of the O2-tolerant [NiFe]-hydrogenase 1 from *Escherichia coli*. Proceedings of the National Academy of Sciences of the United States of America. 2012;**109**(14): 5305-5310

[35] Chengteh Lee WY, Parr RG. Development of the Colle-Salvetti correlation-energy formula into a functional of the electron density. Physical Review B. 1988;**37**: 785

[36] Schäfer A, Horn H, Ahlrichs R. Fully optimized contracted Gaussian basis sets for atoms Li to Kr. The Journal of Chemical Physics. 1992;**97**(4):2571

[37] Schäfer A, Huber C, Ahlrichs R. Fully optimized contracted Gaussian basis sets of triple zeta valence quality for atoms Li to Kr. The Journal of Chemical Physics. 1994;**100**(8): 5829-5835

[38] McLean AD, Chandler GS. Contracted Gaussian basis sets for molecular calculations. I. Second row atoms, Z=11–18. The Journal of Chemical Physics. 1980;**72**(10): 5639-5648

[39] Beinert H, Holm RH, Munck E. Iron-sulfur clusters: Nature's modular, multipurpose structures. Science. 1997; **277**(5326):653-659

[40] Pelmenschikov V, Kaupp M. Redox-dependent structural transformations of the [4Fe-3S] proximal cluster in O_2-tolerant membrane-bound [NiFe]-hydrogenase: A DFT study. Journal of the American Chemical Society. 2013; **135**(32):11809-11823

[41] Shafaat HS, Rudiger O, Ogata H, Lubitz W. [NiFe] hydrogenases: A common active site for hydrogen metabolism under diverse conditions. Biochimica et Biophysica Acta. 2013; **1827**(8–9):986-1002

[42] Volbeda A, Darnault C, Parkin A, Sargent F, Armstrong FA, Fontecilla-Camps JC. Crystal structure of the O_2-tolerant membrane-bound hydrogenase 1 from *Escherichia coli* in complex with its cognate cytochrome b. Structure. 2013;**21**(1):184-190

[43] Kurkin S, George SJ, Thorneley RN, Albracht SP. Hydrogen-induced activation of the [NiFe]-hydrogenase from *Allochromatium vinosum* as studied by stopped-flow infrared spectroscopy. Biochemistry. 2004; **43**(21):6820-6831

[44] Balabin IA, Hu X, Beratan DN. Exploring biological electron transfer pathway dynamics with the pathways plugin for VMD. Journal of Computational Chemistry. 2012;**33**(8): 906-910

[45] Beratan DN, Betts JN, Onuchic JN. Protein electron transfer rates set by the bridging secondary and tertiary structure. Science. 1991;**252**(5010): 1285-1288

[46] Hagiwara Y, Kino H, Tateno M. Modulation of electronic structures of bases through DNA recognition of protein. Journal of Physics. Condensed Matter. 2010;**22**(15):152101

[47] Hagiwara Y, Ohta T, Tateno M. QM/MM hybrid calculation of biological macromolecules using a new interface program connecting QM and MM engines. Journal of Physics. Condensed Matter. 2009;**21**(6):064234

[48] Berg JM, Tymoczko JL, Stryer L. Biochemistry. 5th ed. New York: W. H. Freeman; 2002. p. 1208

[49] Cusack S, Berthet-Colominas C, Hartlein M, Nassar N, Leberman R. A second class of synthetase structure revealed by X-ray analysis of *Escherichia*

coli seryl-tRNA synthetase at 2.5 Å. Nature. 1990;**347**(6290):249-255

[50] Eriani G, Delarue M, Poch O, Gangloff J, Moras D. Partition of tRNA synthetases into two classes based on mutually exclusive sets of sequence motifs. Nature. 1990;**347**(6289):203-206

[51] Delarue M, Moras D. The aminoacyl-tRNA synthetase family: Modules at work. BioEssays. 1993; **15**(10):675-687

[52] Ibba M, Morgan S, Curnow AW, Pridmore DR, Vothknecht UC, Gardner W, et al. A euryarchaeal lysyl-tRNA synthetase: Resemblance to class I synthetases. Science. 1997;**278**(5340): 1119-1122

[53] Fukai S, Nureki O, Sekine S, Shimada A, Tao J, Vassylyev DG, et al. Structural basis for double-sieve discrimination of L-valine from L-isoleucine and L-threonine by the complex of $tRNA^{Val}$ and valyl-tRNA synthetase. Cell. 2000;**103**(5):793-803

[54] Fukunaga R, Fukai S, Ishitani R, Nureki O, Yokoyama S. Crystal structures of the CP1 domain from *Thermus thermophilus* isoleucyl-tRNA synthetase and its complex with L-valine. The Journal of Biological Chemistry. 2004;**279**(9):8396-8402

[55] Fukunaga R, Yokoyama S. Structural basis for non-cognate amino acid discrimination by the valyl-tRNA synthetase editing domain. The Journal of Biological Chemistry. 2005;**280**(33): 29937-29945

[56] Lincecum TL Jr, Tukalo M, Yaremchuk A, Mursinna RS, Williams AM, Sproat BS, et al. Structural and mechanistic basis of pre- and posttransfer editing by leucyl-tRNA synthetase. Molecular Cell. 2003;**11**(4):951-963

[57] Mursinna RS, Lincecum TL Jr, Martinis SA. A conserved threonine within *Escherichia coli* leucyl-tRNA synthetase prevents hydrolytic editing of leucyl-$tRNA^{Leu}$. Biochemistry. 2001; **40**(18):5376-5381

[58] Nureki O, Vassylyev DG, Tateno M, Shimada A, Nakama T, Fukai S, et al. Enzyme structure with two catalytic sites for double-sieve selection of substrate. Science. 1998;**280**(5363): 578-582

[59] Silvian LF, Wang J, Steitz TA. Insights into editing from an ile-tRNA synthetase structure with $tRNA^{ile}$ and mupirocin. Science. 1999;**285**(5430): 1074-1077

[60] Zhai Y, Martinis SA. Two conserved threonines collaborate in the *Escherichia coli* leucyl-tRNA synthetase amino acid editing mechanism. Biochemistry. 2005; **44**(47):15437-15443

[61] Schmidt MW, Baldridge KK, Boatz JA, Elbert ST, Gordon MS, Jensen JH, et al. General Atomic and Molecular Electronic-Structure System. Journal of Computational Chemistry. 1993;**14**(11): 1347-1363

[62] Case DA, Cheatham TE, Darden T, Gohlke H, Luo R, Merz KM, et al. The Amber biomolecular simulation programs. Journal of Computational Chemistry. 2005;**26**(16):1668-1688

[63] Hagiwara Y, Field MJ, Nureki O, Tateno M. Editing mechanism of aminoacyl-trna synthetases operates by a hybrid ribozyme/protein catalyst. Journal of the American Chemical Society. 2010;**132**(8):2751-2758

[64] Cech TR. Crawling out of the RNA world. Cell. 2009;**136**(4):599-602

[65] Kang J, Hagiwara Y, Tateno M. Biological applications of hybrid quantum mechanics/molecular mechanics calculation. Journal of Biomedicine & Biotechnology. 2012; **2012**:11

[66] Kang J, Kino H, Field MJ, Tateno M. Electronic structure rearrangements in hybrid ribozyme/protein catalysis. Journal of the Physical Society of Japan. 2017;**86**(4):044801

[67] Weinger JS, Parnell KM, Dorner S, Green R, Strobel SA. Substrate-assisted catalysis of peptide bond formation by the ribosome. Nature Structural and Molecular Biology. 2004;**11**:1101

[68] Dulic M, Cvetesic N, Zivkovic I, Palencia A, Cusack S, Bertosa B, et al. Kinetic origin of substrate specificity in post-transfer editing by leucyl-tRNA synthetase. Journal of Molecular Biology. 2018;**430**(1):1-16

[69] Nordin BE, Schimmel P. Transiently misacylated tRNA is a primer for editing of misactivated adenylates by class I aminoacyl-tRNA synthetases. Biochemistry. 2003;**42**(44):12989-12997

[70] Nordin BE, Schimmel P. Plasticity of recognition of the 3′-end of mischarged tRNA by class I aminoacyl-tRNA Synthetases. The Journal of Biological Chemistry. 2002;**277**(23): 20510-20517

[71] Kumar S, Das M, Hadad CM, Musier-Forsyth K. Substrate and enzyme functional groups contribute to translational quality control by bacterial prolyl-tRNA synthetase. The Journal of Physical Chemistry B. 2012;**116**(23): 6991-6999

[72] Waas WF, Schimmel P. Evidence that tRNA synthetase-directed proton transfer stops mistranslation. Biochemistry. 2007;**46**(43): 12062-12070

[73] Aboelnga MM, Hayward JJ, Gauld JW. Enzymatic post-transfer editing mechanism of *E. coli* threonyl-tRNA synthetase (ThrRS): A molecular dynamics (MD) and quantum mechanics/molecular mechanics (QM/MM) investigation. ACS Catalysis. 2017; **7**(8):5180-5193

[74] Hussain T, Kruparani SP, Pal B, Dock-Bregeon AC, Dwivedi S, Shekar MR, et al. Post-transfer editing mechanism of a D-aminoacyl-tRNA deacylase-like domain in threonyl-tRNA synthetase from archaea. The EMBO Journal. 2006;**25**(17):4152-4162

[75] Aboelnga MM, Hayward JJ, Gauld JW. Unraveling the critical role played by Ado762′OH in the post-transfer editing by archaeal threonyl-tRNA synthetase. The Journal of Physical Chemistry B. 2018;**122**(3):1092-1101

[76] Ling J, Roy H, Ibba M. Mechanism of tRNA-dependent editing in translational quality control. Proceedings of the National Academy of Sciences. 2007;**104**(1):72

[77] Fukunaga R, Yokoyama S. Structural basis for substrate recognition by the editing domain of isoleucyl-tRNA synthetase. Journal of Molecular Biology. 2006;**359**(4):901-912

[78] Aboelnga MM, Hayward JJ, Gauld JW. A water-mediated and substrate-assisted aminoacylation mechanism in the discriminating aminoacyl-tRNA synthetase GlnRS and non-discriminating GluRS. Physical Chemistry Chemical Physics. 2017; **19**(37):25598-25609

[79] Fukui K. Recognition of Stereochemical paths by orbital interaction. Accounts of Chemical Research. 1971;**4**(2):57

Section 3

Structural Quantum Developments

Chapter 5

Supersymmetric Quantum Mechanics: Two Factorization Schemes and Quasi-Exactly Solvable Potentials

José Socorro García Díaz, Marco A. Reyes, Carlos Villaseñor Mora and Edgar Condori Pozo

Abstract

We present the general ideas on supersymmetric quantum mechanics (SUSY-QM) using different representations for the operators in question, which are defined by the corresponding bosonic Hamiltonian as part of SUSY Hamiltonian and its supercharges, which are defined as matrix or differential operators. We show that, although most of the SUSY partners of one-dimensional Schrödinger problems have already been found, there are still some unveiled aspects of the factorization procedure which may lead to richer insights of the problem involved.

Keywords: supersymmetric quantum mechanics, quasi-exactly solvable potentials

1. Introduction

We present the general ideas on supersymmetric quantum mechanics (SUSY-QM) using different representations for the operators in question, which are defined by the corresponding bosonic Hamiltonian as part of SUSY Hamiltonian and its supercharges, $\hat{Q}^-$ and $\bar{Q}^+$, which are defined as matrix or differential operators. We show that, although most of the SUSY partners of one-dimensional Schrödinger problems have already been found [1], there are still some unveiled aspects of the factorization procedure which may lead to richer insights of the problem involved. In particular, we refer to the factorization of the Hamiltonian in terms of two non-mutually adjoint operators [2, 3].

In this work, we try three main schemes; the first one consists on finding the eigenvalue Schrödinger equation in one dimension using the matrix representation via the appropriate factorization with ladder-like operators and finding the one parameter isospectral equation for this one. In this scheme, the wave function is written as a supermultiplet. Continuing with the Schrödinger model, we extend SUSY to include two-parameter factorizations, which include the SUSY factorization as particular case. As examples, we include the case of the harmonic oscillator and the Pöschl-Teller potentials. Also, we include the steps for the two-dimensional case and apply it to particular cases. The second scheme uses the differential representation in Grassmann numbers, where the wave function can be written as an n-dimensional vector or as an expansion in Grassmann variables multiplied by bosonic functions. We apply the scheme in two bosonic variables a particular cosmological model and compare the corresponding solutions found. The third scheme tries on extensions to the SUSY factorization and to the case of quasi-exactly solvable potentials; we present a particular case which does not form part of the class of potentials found using Lie algebras.

To establish the different approaches presented here, we will briefly describe the different main formalisms applied to supersymmetric quantum mechanics, techniques that are now widely used in a rich spectrum of physical problems, covering such diverse fields as particle physics, quantum field theory, quantum gravity, quantum cosmology, and statistical mechanics, to mention some of them:

- In one dimension, SUSY-QM may be considered an equivalent formulation of the Darboux transformation method, which is well-known in mathematics from the original paper of Darboux [4], the book by Ince [5], and the book by Matweev and Salle [6], where the method is widely used in the context of the soliton theory. An essential ingredient of the method is the particular choice of a transformation operator in the form of a differential operator which intertwines two Hamiltonian and relates their eigenfunctions. When this approach is applied to quantum theory, it allows to generate a huge family of exactly solvable local potential starting with a given exactly solvable local potential [7]. This technique is also known in the literature as isospectral formalism [7–10].

- Those defined by means of the use of supersymmetry as a square root [11–14], in which the Grassmann variables are auxiliary variables and are not identified as the supersymmetric partners of the bosonic variables. In this formalism, a differential representation is used for the Grassmann variables. Also the supercharges for the n-dimensional case read as

$$\hat{Q}^{-} = \psi^{\mu}\left[-\hbar\partial_{q^{\mu}} + \frac{\partial S}{\partial q^{\mu}}\right], \qquad \hat{Q}^{+} = \overline{\psi}^{\nu}\left[-\hbar\partial_{q^{\nu}} - \frac{\partial S}{\partial q^{\nu}}\right], \tag{1}$$

where S is known as the superpotential function which are related to the physical potential under consideration, when the Hamiltonian density is written as the Hamilton-Jacobi equation, and the algebra for the variables ψ^{μ} and $\overline{\psi}^{\nu}$ is

$$\{\psi^{\mu}, \overline{\psi}^{\nu}\} = \eta^{\mu\nu}, \qquad \{\psi^{\mu}, \psi^{\nu}\} = 0, \qquad \{\overline{\psi}^{\mu}, \overline{\psi}^{\nu}\} = 0. \tag{2}$$

There are two forms where the equations in 1D are satisfied: in the literature we find either the matrix representation or the differential operator scheme. However for more than one dimensions, there exist many applications to cosmological models, where the differential representation for the Grassmann variables is widely applied [14–18]. There are few works in more dimensions in the first scheme [19]; we present in this work the main ideas to build the 2D case, where the supercharge operators become 4×4 matrices.

2. Factorization method in one dimension: matrix approach

We begin by introducing the main ideas for the one-dimensional quantum harmonic oscillator. The corresponding Hamiltonian is written in operator form as

$$\hat{\mathrm{H}}_{\mathrm{B}} = \frac{1}{2}\hat{\mathrm{p}}^2 + \frac{1}{2}\omega_{\mathrm{B}}^2\hat{\mathrm{q}}^2 \tag{3}$$

where $\hat{\mathrm{q}}$ is the generalized coordinate and $\hat{\mathrm{p}}$ is the associated momentum, the canonical commutation relation between this quantities being $[\hat{\mathrm{q}}, \hat{\mathrm{p}}] = \mathrm{i}$. We introduce two new operators, known as the creation and annihilation operators $\hat{\mathrm{a}}^+$, $\hat{\mathrm{a}}^-$, respectively, defined as

$$\hat{a}^- = \frac{1}{\sqrt{2\omega_B}}(\hat{p} - i\omega_B\hat{q}), \qquad \hat{a}^+ = \frac{1}{\sqrt{2\omega_B}}(\hat{p} + i\omega_B\hat{q}). \tag{4}$$

This Hamiltonian can be written in terms of the anti-commutation relation between these operators as

$$\hat{H}_B = \frac{\omega_B}{2}\{\hat{a}^+, \hat{a}^-\}. \tag{5}$$

The symmetric nature of $\hat{\mathrm{H}}_{\mathrm{B}}$ under the interchange of $\hat{a}^-$ and $\hat{a}^+$ suggests that these operators satisfy Bose-Einstein statistics, and it is therefore called bosonic.

Now, we build the operators $\hat{b}^-$ and $\hat{b}^+$ that obey similar rules to operators $\hat{a}^-, \hat{a}^+$ changing $[,] \leftrightarrows \{,\}$, that is,

$$\left\{\hat{b}^-, \hat{b}^+\right\} = 1; \qquad \left\{\hat{b}^-, \hat{b}^-\right\} = \left\{\hat{b}^+, \hat{b}^+\right\} = 0, \tag{6}$$

and in analogy to (5), we define the corresponding new Hamiltonian as

$$\hat{\mathrm{H}}_{\mathrm{F}} = \frac{\omega_{\mathrm{F}}}{2}\left[\hat{\mathrm{b}}^+, \hat{\mathrm{b}}^-\right]. \tag{7}$$

The antisymmetric nature of $\hat{\mathrm{H}}_{\mathrm{F}}$ under the interchange of $\hat{b}^-$ and $\hat{b}^+$ suggests that these operators satisfy the Fermi-Dirac statistics, and it is called fermionic.

These operators $\hat{b}^-$ and $\hat{b}^+$ admit a matrix representations in terms of Pauli matrices that satisfy all rules defined above, that is,

$$\hat{b}^- = \sigma_-, \qquad \hat{b}^+ = \sigma_+, \qquad \sigma_\pm = \frac{1}{2}(\sigma_1 \pm i\sigma_2) \tag{8}$$

with $[\sigma_+, \sigma_-] = \sigma_3$, $\sigma_- = \begin{pmatrix} 0 & 0 \\ 1 & 0 \end{pmatrix}$, $\sigma_+ = \begin{pmatrix} 0 & 1 \\ 0 & 0 \end{pmatrix}$, $\sigma_1 = \begin{pmatrix} 0 & 1 \\ 1 & 0 \end{pmatrix}$, $\sigma_2 = \begin{pmatrix} 0 & -i \\ i & 0 \end{pmatrix}$, $\sigma_3 = \begin{pmatrix} 1 & 0 \\ 0 & -1 \end{pmatrix}$.

Now, consider both Hamiltonians as a composite system, that is, we consider the superposition of two oscillators, one being bosonic and one fermionic, with energy $\mathrm{E_T} = \mathrm{E_B} + \mathrm{E_F}$

$$\mathrm{E_T} = \omega_{\mathrm{B}}\left(\mathrm{n_B} + \frac{1}{2}\right) + \omega_{\mathrm{F}}\left(\mathrm{n_F} - \frac{1}{2}\right) = \omega_{\mathrm{B}}\mathrm{n_B} + \omega_{\mathrm{F}}\mathrm{n_F} + \frac{1}{2}(\omega_{\mathrm{B}} - \omega_{\mathrm{F}}). \tag{9}$$

When we demand that both frequencies are the same, $\omega_B = \omega_F = \omega$, we introduce a new symmetry, called supersymmetry (SUSY); we can see that the simultaneous creation of a quantum fermion $(n_F \to n_F + 1)$ causes the destruction of quantum boson $(n_B \to n_B - 1)$ and vice versa, in the sense that the total energy is unaltered. The ground energy state is exact and no degenerate. The degeneration appears from n = 1, where it is double degenerate.

In this way, we have the super-Hamiltonian $\hat{H}_{susy}$, written as

$$\hat{H}_{susy} = \frac{\omega}{2}\{\hat{a}^+,\hat{a}^-\} + \frac{\omega}{2}\left[\hat{b}^+,\hat{b}^-\right] = \frac{\omega}{2}\{\hat{a}^+,\hat{a}^-\}I + \frac{\omega}{2}\sigma_3 = \omega\begin{pmatrix} \hat{a}-\hat{a}^+ & 0 \\ 0 & \hat{a}^+\hat{a}^- \end{pmatrix}$$
$$= \begin{pmatrix} \hat{H}_- & 0 \\ 0 & \hat{H}_+ \end{pmatrix}, \tag{10}$$

where I is a 2×2 unit matrix and where the two components of $\hat{H}_{susy}$ in (10) can be written independently as

$$\hat{H}_+ = \frac{1}{2}\hat{p}^2 + \frac{1}{2}\left(\omega^2 q^2 - \omega\right) \equiv \omega\hat{a}^+\hat{a}^- \tag{11}$$

$$\hat{H}_- = \frac{1}{2}\hat{p}^2 + \frac{1}{2}\left(\omega^2 q^2 + \omega\right) \equiv \omega\hat{a}^-\hat{a}^+. \tag{12}$$

From Eqs. (18) and (19), we can see that $\hat{H}_+$ and $\hat{H}_-$ are the same representation of one Hamiltonian with a constant shifting ω in the energy spectrum.

The question is, what are the generators for this SUSY Hamiltonian? The answer is, considering that the degeneration is the result of the simultaneous destruction (creation) of quantum boson and the creation (destruction) of quantum fermion, the corresponding generators for this symmetry must be written as $\hat{a}^-\hat{b}^+$ (or $\hat{a}^+\hat{b}^-$). Therefore we introduce the following generators, called supercharges $\hat{Q}^-$ and $\hat{Q}^+$, defined as

$$\hat{Q}^- = \sqrt{2\omega}\hat{a}^-\hat{b}^+ = \sqrt{2\omega}\begin{pmatrix} 0 & \hat{a}^- \\ 0 & 0 \end{pmatrix}, \qquad \hat{Q}^+ = \sqrt{2\omega}\hat{a}^+\hat{b}^- = \sqrt{2\omega}\begin{pmatrix} 0 & 0 \\ \hat{a}^+ & 0 \end{pmatrix}, \tag{13}$$

implying that

$$\hat{H}_{susy} = \frac{1}{2}\left\{\hat{Q}^+,\hat{Q}^-\right\} \tag{14}$$

and satisfying the following relations

$$\left\{\hat{Q}^-,\hat{Q}^-\right\} = \left\{\hat{Q}^+,\hat{Q}^+\right\} = 0; \quad \left[\hat{Q}^-,\hat{H}_{susy}\right] = \left[\hat{Q}^+,\hat{H}_{susy}\right] = 0. \tag{15}$$

We can generalize this procedure for a certain function W(q), and at this point, we can define two new operators $\hat{A}^-$ and $\hat{A}^+$ with a property similar to (4),

$$\hat{A}^- = \frac{1}{\sqrt{2\omega}}(\hat{p} - i\omega W(q)), \qquad \hat{A}^+ = \frac{1}{\sqrt{2\omega}}(\hat{p} + i\omega W(q)). \tag{16}$$

In order to obtain the general solutions, we can use an arbitrary potential in Eq. (3), that is,

$$\hat{H}_B = \frac{1}{2}\hat{p}^2 + V(q). \tag{17}$$

The Hamiltonians $\hat{H}^+$ and $\hat{H}^-$ determine two new potentials,

$$\hat{H}_+ = \frac{1}{2}\hat{p}^2 + V_+ = \frac{1}{2}\hat{p}^2 + \frac{1}{2}\left(W^2 - \frac{dW}{dq}\right) \tag{18}$$

$$\hat{H}_- = \frac{1}{2}\hat{p}^2 + V_- = \frac{1}{2}\hat{p}^2 + \frac{1}{2}\left(W^2 + \frac{dW}{dq}\right), \tag{19}$$

where the potential term $V_+(q)$ is related to the superpotential function $W(q)$ via the Riccati equation

$$V_+ = \frac{1}{2}\left(W^2 - \frac{dW}{dq}\right), \tag{20}$$

(modulo constant ϵ, which is related to some energy eigenvalue) and $V_- = \frac{1}{2}\left(W^2 + \frac{dW}{dq}\right) = V_+ + \frac{dW}{dq}$, with the same spectrum, except for the ground state, which is related to the energy potential V_+.

In a general way, let us now find the general form of the function W. The quantum equation (17) applied to stationary wave function u_i becomes

$$-\frac{1}{2}\frac{d^2u_i}{dq^2} + V(q)u_i = E_iu_i, \tag{21}$$

where E_i are the energy eigenvalues. Considering the transformation $W(q) = -\frac{d\ln[u_i(q)]}{dq}$ and introducing it into (18), we have that

$$V(q) - E_i = \frac{1}{2}\left(W^2 - \frac{dW}{dq}\right) = \left(\frac{1}{2u_i}\frac{du_i}{dq}\right)^2 - \frac{\left(\frac{du_i}{dq}\right)^2 - u_i\frac{d^2u_i}{dq^2}}{2u_i^2} = \frac{1}{2u_i}\frac{d^2u_i}{dq^2}.$$

Then, this equation is the same as the original one, Eq. (21), that is, W is related to an initial solution of the bosonic Hamiltonian. What happens to the isopotential $V_-(q) = \frac{1}{2}\left(W^2 + \frac{dW}{dq}\right)$? Considering that

$$2V_- = W^2 + \frac{dW}{dq} \equiv \hat{W}^2 + \frac{d\hat{W}}{dq} = 2\hat{V}_-,$$

the question is, what is $\hat{W}$ if we know the function W? Finding it we can build a family of potentials $\hat{V}_-$ depending on a free parameter λ, the supersymmetric parameter that, to some extent, plays the role of internal time. Following the procedure $\hat{W} = W + \frac{1}{y(q)}$, where the function $y(q)$ satisfy the linear differential equation $\frac{dy}{dq} - 2Wy = 1$, the solution implies

$$y(q) = \frac{\lambda + \int u_i^2 dq}{u_i^2}, \quad \rightarrow \quad \hat{W} = W + \frac{u_i^2}{\lambda + \int u_i^2 dq}. \tag{22}$$

The family of potentials $\hat{V}_+$ can be built now as

$$\hat{V}_+ - E_i = \frac{1}{2}\left(\hat{W}^2 - \frac{d\hat{W}}{dq}\right) = V_- + \frac{d\hat{W}}{dq}. \tag{23}$$

Finally

$$\hat{u} = g(\lambda)\frac{u_i}{\lambda + \int u_i^2 dq} \tag{24}$$

is the isospectral solution of the Schrödinger-like equation for the new family potential (23), with the condition $g(\lambda) = \sqrt{\lambda(\lambda+1)}$, which in the limit

$$\lambda \to \pm\infty, \qquad g(\lambda) = \lambda, \qquad \hat{u}_i \to u_i.$$

This λ parameter is included not for factorization reasons; in particular, in quantum cosmology the wave functions are still nonnormalizable, and λ is used as a decoherence parameter embodying a sort of quantum cosmological dissipation (or damping) distance.

2.1 Two-dimensional case

We use Witten's idea [20] to find the supersymmetric supercharge operators Q^- and Q^+ that generate the super-Hamiltonian H_{susy}. Using Eqs. (13)–(15), we can generalize the one-dimensional factorization scheme. We define the two-dimensional Hamiltonian as

$$\hat{H}_B(x, y) = \frac{1}{2}\hat{p}_x^2 + \frac{1}{2}\hat{p}_y^2 + V(x) + V(y), \tag{25}$$

where the Schrödinger-like equation can be obtained as the bosonic sector of this super-Hamiltonian in the superspace, i.e., when all fermionic fields are set equal to zero (classical limit).

In two dimensions, the supercharges are defined by the tensorial products

$$Q^- = \sqrt{2}d\Big|^- \otimes \sigma_+, \qquad Q^+ = \sqrt{2}d\Big|^+ \otimes \sigma_- \tag{26}$$

with

$$d\Big|^- = \begin{pmatrix} a^- & 0 \\ 0 & b^- \end{pmatrix}, \qquad d\Big|^+ = \begin{pmatrix} a^+ & 0 \\ 0 & b^+ \end{pmatrix}, \tag{27}$$

where $\sigma_\pm$ are the same as in (8). From Eq. (26) we have that the supercharges are 4×4 matrices

$$\hat{Q}^+ = \sqrt{2}\begin{bmatrix} 0 & 0 & 0 & 0 \\ 0 & 0 & 0 & 0 \\ a^+ & 0 & 0 & 0 \\ 0 & b^+ & 0 & 0 \end{bmatrix} \qquad \hat{Q}^- = \sqrt{2}\begin{bmatrix} 0 & 0 & a^- & 0 \\ 0 & 0 & 0 & b^- \\ 0 & 0 & 0 & 0 \\ 0 & 0 & 0 & 0 \end{bmatrix} \tag{28}$$

where the super-Hamiltonian, (14), can be written as

$$H_{susy} = \begin{pmatrix} a^-a^+ & 0 & 0 & 0 \\ 0 & b^-b^+ & 0 & 0 \\ 0 & 0 & a^+a^- & 0 \\ 0 & 0 & 0 & b^+b^- \end{pmatrix} = \begin{pmatrix} H_-^1(x) & 0 & 0 & 0 \\ 0 & H_-^1(y) & 0 & 0 \\ 0 & 0 & H_+^2(x) & 0 \\ 0 & 0 & 0 & H_+^2(y) \end{pmatrix}, \tag{29}$$

where

$$a^- = \frac{1}{\sqrt{2}}\left(\frac{d}{dx} + W(x)\right), \qquad a^+ = \frac{1}{\sqrt{2}}\left(-\frac{d}{dx} + W(x)\right) \tag{30}$$

$$b^- = \frac{1}{\sqrt{2}}\left(\frac{d}{dx} + Z(y)\right), \qquad b^+ = \frac{1}{\sqrt{2}}\left(-\frac{d}{dx} + Z(y)\right) \tag{31}$$

and $V(x,y) = W(x) + Z(y)$.
The Riccati equation (20) is written in 2D as

$$V_+(x,y) = V_+1(x) + V_{+2}(y) = \frac{1}{2}\left(W^2 - \frac{dW}{dx}\right) + \frac{1}{2}\left(Z^2 - \frac{dZ}{dy}\right), \tag{32}$$

and, using separation variables, we get

$$V_1(x) - \frac{1}{2}\left(W^2(x) - \frac{dW}{dx}\right) = C_0 \tag{33}$$

$$V_2(y) - \frac{1}{2}\left(Z^2(y) - \frac{dZ}{dy}\right) = -C_0. \tag{34}$$

In general, we find that each potential V_{+i} satisfies

$$\frac{1}{2}\frac{d^2}{dx^2}u_i(x) + V_{+i}u_i(x) = E_i u_i(x), \qquad i = 1, 2, \tag{35}$$

and we can find the isopotential as $W = -\frac{1}{u_1}\frac{du_1}{dx}$, when u_1 is known.

Following the same steps as in the 1D case, we find that the solutions (22) are the same in this case. So, the general solution for $\hat{W}$ is $\hat{W} = W + \frac{1}{y(x)}$, with $y = u_1^{-2}(x)\left[E_1 + \int u_1^2(x)dx\right]$. The general solution for the superpotential $\hat{W}(x)$ is

$$\hat{W} = -\frac{1}{u_1}\frac{du_1}{dx} + \frac{u_1^2}{\lambda_1 + \int u_1^2 dx} = W_p + \frac{d}{dx}[\mathrm{Ln}(\lambda_1 + I_1)] \tag{36}$$

where $W_p = -\frac{1}{u_1}\frac{du_1}{dx}$ and $I_1 = \int u_1^2 dx$.
In the same manner, we have that

$$\hat{Z} = -\frac{1}{u_2}\frac{du_2}{dy} + \frac{u_2^2}{\lambda_2 + \int u_2^2 dy} = Z_p + \frac{d}{dy}[\mathrm{Ln}(\lambda_2 + I_2)] \tag{37}$$

with $Z_p = -\frac{1}{u_2}\frac{su_2}{dy}$ and $I_2 = \int u_2^2 dy$.

On the other hand, using the Riccati equation, we can build a generalization for the isopotential, using the new potential $\hat{W}$, as

$$\hat{V}_{+1}(x,\lambda_1) = \frac{1}{2}\left(\hat{W}^2 - \hat{W}'\right) = V_+(x) - \frac{2u_1\frac{du_1}{dx}}{\lambda_1 + I_1} + \frac{u_1^4}{(\lambda_1 + I_1)^2}. \tag{38}$$

For the other coordinate, we have

$$\hat{V}_{+2}(y,\lambda_2) = \frac{1}{2}\left(\hat{Z}^2 - \frac{d\hat{Z}}{dy}\right) = V_+(y) - \frac{2u_2\frac{du_2}{dy}}{\lambda_2 + I_2} + \frac{u_2^4}{(\lambda_2 + I_2)^2}. \tag{39}$$

The general solutions for $\hat{u}_i$ depend on the initial solutions to the original Schrödinger equations in the variables (x,y), that is, $u_1 = u_1(x)$, $u_2 = u_2(y)$, being

$$\hat{u}_1(x,\lambda_1) = C_1(\lambda_1)\frac{u_1}{\lambda_1 + I_1}, \qquad \hat{u}_2(y,\lambda_2) = C_2(\lambda_2)\frac{u_2}{\lambda_2 + I_2}, \tag{40}$$

where the *variables* $C_i(\lambda_i)$ have the same properties that $g(\lambda)$ obtained in the 1D case.

2.2 Application to cosmological Taub model

The Wheeler-DeWitt equation for the cosmological Taub model is given by

$$\frac{\partial^2\Psi}{\partial\alpha^2} - \frac{\partial^2\Psi}{\partial\beta^2} + e^{4\alpha}V(\beta)\Psi = 0 \tag{41}$$

where $V(\beta) = \frac{1}{3}\left(e^{-8\beta} - 4e^{-2\beta}\right)$. These equations can be separated using $x_1 = 4\alpha - 8\beta$ and $x_2 = 4\alpha - 2\beta$, rendering

$$-\frac{\partial^2 f_1(x_1)}{\partial x_1^2} + \frac{1}{144}e^{x_1}f_1(x_1) = \frac{\omega^2}{4}f_1(x_1), \qquad -\frac{\partial^2 f_2(x_2)}{\partial x_2^2} + \frac{1}{9}e^{x_2}f_2(x_2) = \omega^2 f_2(x_2), \tag{42}$$

where the parameter ω is the separation constant. These equations possess the solutions

$$f_1 = K_{i\omega}\left(\frac{1}{6}e^{\frac{x_1}{2}}\right), \qquad f_2 = L_{2i\omega}\left(\frac{2}{3}e^{\frac{x_2}{2}}\right) + K_{2i\omega}\left(\frac{2}{3}e^{\frac{x_2}{2}}\right) \tag{43}$$

where K (or I) is the modified Bessel function of imaginary order and the function L is defined as

$$L_{2i\omega} = \frac{\pi i}{2\sinh(2\omega\pi)}\left(I_{2i\omega} + I_{-2i\omega}\right).$$

Using Eqs. (38) and (39), we obtain the isopotential for this model

$$\begin{aligned}\hat{V}(x_1) &= V_+(x_1) - \frac{2K_{i\omega}K'_{i\omega}}{\lambda_1 + I_1} + \frac{K_{i\omega}^4}{(\lambda_1 + I_1)^2},\\ \hat{V}(x_2) &= V_+(x_2) - \frac{2(L_{2i\omega} + K_{2i\omega})(L_{2i\omega} + K_{2i\omega})'}{\lambda_2 + I_2} + \frac{(L_{2i\omega} + K_{2i\omega})^4}{(\lambda_2 - I_2)^2}.\end{aligned} \tag{44}$$

Using Eq. (40) we can obtain general solutions for the functions f_1 and f_2 in the following way

$$\hat{f}_1 = \frac{C_1 K_{i\omega}\left(\frac{1}{6}e^{\frac{x_1}{2}}\right)}{\lambda_1 + I_1}, \qquad \hat{f}_2 = \frac{C_2\left[L_{2i\omega}\left(\frac{2}{3}e^{\frac{x_2}{2}}\right) + K_{2i\omega}\left(\frac{2}{3}e^{\frac{x_2}{2}}\right)\right]}{\lambda_2 + I_2}. \tag{45}$$

3. Differential approach: Grassmann variables

The supersymmetric scheme has the particularity of being very restrictive, because there are many constraint equations applied to the wave function. So, in this work and in others, we found that there exist a tendency for supersymmetric vacua to remain close to their semiclassical limits, because the exact solutions found are also the lowest-order WKB-like approximations and do not correspond to the full quantum solutions found previously for particular models [14–18].

Maintaining the structure of Eqs. (13)–(16), taking the differential representation for the fermionic operator $\hat{b} \leftrightarrow \psi^\mu$ for convenience in the calculations, and changing the function $W \to \frac{\partial S}{\partial q^\mu}$, the supercharges for the n-dimensional case read as

$$\hat{Q}^- = \psi^\mu \left[P_\mu + i\frac{\partial S}{\partial q^\mu}\right], \qquad \hat{Q}^+ = \overline{\psi}^\nu \left[P_\nu - i\frac{\partial S}{\partial q^\nu}\right], \tag{46}$$

where S is known as the superpotential functions which are related to the physical potential under consideration, when the Hamiltonian density is written as the Hamilton-Jacobi equation, and the following algebra for the variables ψ^μ and $\overline{\psi}^\nu$ (similar to Eq. (6))

$$\{\psi^\mu, \overline{\psi}^\nu\} = \eta^{\mu\nu}, \qquad \{\psi^\mu, \psi^\nu\} = 0, \qquad \{\overline{\psi}^\mu, \overline{\psi}^\nu\} = 0. \tag{47}$$

These rules are satisfied when we use a differential representation for these $\psi^\mu, \overline{\psi}^\nu$ variables in terms of the Grassmann numbers, as

$$\psi^\mu = \eta^{\mu\nu}\frac{\partial}{\partial\theta^\nu}, \qquad \overline{\psi}^\nu = \theta^\nu, \tag{48}$$

where $\eta^{\mu\nu}$ is a diagonal constant matrix, its dimensions depending on the independent bosonic variables that appear in the bosonic Hamiltonian. Now the super-Hamiltonian is written as

$$H_S = \frac{1}{2}\left\{\hat{Q}^+, \hat{Q}^-\right\} = \mathcal{H}_0 + \frac{\hbar}{2}\frac{\partial^2 S}{\partial q^\mu \partial q^\nu}[\overline{\psi}^\mu, \psi^\nu], \tag{49}$$

where $\mathcal{H}_0 = \Box + U(q^\mu)$ is the quantum version of the classical bosonic Hamiltonian, $\Box$ is the d'Alembertian in three dimension when we have three bosonic independent coordinates, and $U(q^\mu)$ is the potential energy in consideration.

The superspace for three-dimensional model becomes $(q_1, q_2, q_3, \theta^0, \theta^1, \theta^2)$, where the variables θ^i are the coordinate in the fermionic space, as the Grassmann numbers, which have the property of $\theta^i\theta^j = -\theta^j\theta^i$, and the wave function has the representation

$$\Psi = \mathcal{A}_+ + \mathcal{B}_0\theta^0, \qquad 1 \text{ dimension} \tag{50}$$

$$\Psi = \mathcal{A}_+ + \mathcal{B}_0\theta^0 + \mathcal{B}_1\theta^1 + \mathcal{A}_-\theta^0\theta^1, \qquad 2 \text{ dimensions} \tag{51}$$

$$\Psi = \mathcal{A}_+ + \mathcal{B}_\nu\theta^\nu + \frac{1}{2}\epsilon_{\mu\nu\lambda}\mathcal{C}^\lambda\theta^\mu\theta^\nu + \mathcal{A}_-\theta^0\theta^1\theta^2, \qquad 3 \text{ dimensions} \tag{52}$$

where the indices μ, ν, λ values are 0, 1, and 2 and $\mathcal{A}_\pm$, $\mathcal{B}_\nu$ and $\mathcal{C}^\lambda$ are bosonic functions which depend on the bosonic coordinates q^μ and not on the Grassmann numbers. Here, the wave function representation structure is set in terms of 2^n

components, for n independent bosonic coordinates, with half of the terms coming from the bosonic (fermionic) contribution into the wave function.

It is well-known that the physical states are determined by the applications of the supercharges $\hat{Q}^{-}$ and $\hat{Q}^{+}$ on the wave functions, that is,

$$\hat{Q}^{-}\Psi = 0, \qquad \hat{Q}^{+}\Psi = 0, \tag{53}$$

where we use the usual representation for the momentum $P_\mu = -i\frac{\partial}{\partial q^\mu}$. Considering the 2D case, the last second equation gives

$$\theta^0 : \left[\frac{\partial A_+}{\partial q^0} - A_+\frac{\partial S}{\partial q^0}\right] = 0, \tag{54}$$

$$\theta^1 : \left[\frac{\partial A_+}{\partial q^1} - A_+\frac{\partial S}{\partial q^1}\right] = 0, \tag{55}$$

$$\theta^0\theta^1 : \left[\frac{\partial B_1}{\partial q^0} - B_1\frac{\partial S}{\partial q^0}\right] - \left[\frac{\partial B_0}{\partial q^1} - B_0\frac{\partial S}{\partial q^1}\right] = 0. \tag{56}$$

From (54)–(55), we obtain the relation $\frac{\partial A_+}{\partial q^\mu} - A_+\frac{\partial S}{\partial q^\mu} = 0$ with the solution $A_+ = a_+ e^S$.

On the other hand, the first equation in (53) gives

$$\theta^0 : \left[\frac{\partial A_-}{\partial q^1} + A_-\frac{\partial S}{\partial q^1}\right] = 0, \tag{57}$$

$$\theta^1 : \left[\frac{\partial A_-}{\partial q^0} + A_-\frac{\partial S}{\partial q^0}\right] = 0, \tag{58}$$

$$\textit{free term} : -\left[\frac{\partial B_0}{\partial q^0} + B_0\frac{\partial S}{\partial q^0}\right] + \left[\frac{\partial B_1}{\partial q^1} + B_1\frac{\partial S}{\partial q^1}\right] = 0. \tag{59}$$

The free term equation is written as $\eta^{\mu\nu}(\partial_\mu B_\nu + B_\nu\partial_\mu S) = 0$, and taking the ansatz $B_\mu = e^{-S}\partial_\nu f_+(q^\mu)$, Eq. (56) is fulfilled, so we obtain for the free term,

$$\Box f_+ + 2\eta^{\mu\nu}\nabla_\mu S\nabla_\nu f_+ = 0, \tag{60}$$

with the solution to $f_+ = h(q^1 - q^2)$, with h an arbitrary function depending of its argument. However, this function f must depend on the potential under consideration.

Also, Eqs. (57) and (58) are written as

$$\frac{\partial A_-}{\partial q^\mu} + A_-\frac{\partial S}{\partial q^\mu} = 0, \qquad \frac{1}{A_-}\frac{\partial A_-}{\partial q^\mu} = -\frac{\partial S}{\partial q^\mu} \qquad \rightarrow \qquad \frac{\partial \mathrm{Ln} A_-}{\partial q^\mu} = -\frac{\partial S}{\partial q^\mu} \tag{61}$$

whose solution is $A_- = a_- e^{-S}$. In this way, all functions entering the wave function are

$$A_\pm = a_\pm e^{\pm S}, \qquad B_0 = e^{-S}\partial_0(f_+), \qquad B_1 = e^{-S}\partial_1(f_+).$$

3.1 The unnormalized probability density

To obtain the wave function probability density $|\Psi|^2$ in this supersymmetric fashion, we need first to integrate over the Grassmann variables θ^i. This procedure is well-known [21], and here we present the main ideas. Let Ψ_1 and Ψ_2 be two functions that depend on Grassmann numbers; the product $<\Psi_1, \Psi_2>$ is defined as

$$<\Psi_1, \Psi_2> = \int (\Psi_1(\theta^*))^* \Psi_2(\theta^*)\, e^{-\sum_i \theta_i^* \theta_i} \Pi_i d\theta_i^* d\theta_i, \qquad (C\theta_i \cdots \theta_r)^* = \theta_r^* \cdots \theta_i^* C^*,$$

and the integral over the Grassmann numbers is $\int \theta_i^* \theta_i \cdots \theta_m \theta_m^* d\theta_m^* d\theta_m \cdots d\theta_i^* d\theta_i = 1$. In 2D, the main contributions to the term $e^{-\sum_i \theta_i^* \theta_i}$ come from

$$e^{-\sum_i \theta_i^* \theta_i} = e^{-\sum_i \theta_i \theta_i^*} = 1 + \theta^0 \theta^{*0} + \theta^1 \theta^{*1} + \theta^0 \theta^{*0} \theta^1 \theta^{*1},$$

and using that $\int \theta d\theta = 1$, and $\int d\theta = 0$, which act as a filter, we obtain that

$$|\Psi|^2 = \mathcal{A}_+^* \mathcal{A}_+ + \mathcal{B}_0^* \mathcal{B}_0 + \mathcal{B}_1^* \mathcal{B}_1 + \mathcal{A}_-^* \mathcal{A}_-.$$

By demanding that $|\Psi|^2$ does not diverge when $|q^0|, |q^1| \to \infty$, only the contribution with the exponential e^{-2S} will remain.

4. Beyond SUSY factorization

Although most of the SUSY partners of 1D Schrödinger problems have been found [1], there are still some unveiled aspects of the factorization procedure. We have shown this for the simple harmonic oscillator in previous works [2, 3] and will proceed here in the same way for the problem of the modified Pöschl-Teller potential. The factorization operators depend on two supersymmetric type parameters, which when the operator product is inverted, allow us to define a new SL operator, which includes the original QM problem.

The Hamiltonian of a particle in a modified Pöschl-Teller potential is [1, 22]

$$H_{m+1}\Psi = \left(-\frac{\hbar^2}{2\mu}\frac{d^2}{dx^2} - \frac{\alpha^2 m(m+1)}{\cosh^2 \alpha x} \right)\Psi = E\Psi\ , \tag{62}$$

where $\alpha > 0$ and the integer m is greater than 0. To shorten the algebraic equations, we shall set $\frac{\hbar^2}{2\mu} = 1$.

The eigenvalue problem may be solved using the Infeld and Hull's (IH) factorizations [23],

$$A_{m+1}^+ A_{m+1}^- \psi_{m-n}^m = (H_{m+1} + \epsilon_{m+1})\psi_{m-n}^m, \tag{63}$$

$$A_m^- A_m^+ \psi_{m-n}^m = (H_{m+1} + \epsilon_m)\psi_{m-n}^m, \tag{64}$$

where the IH raising/lowering operators are given by

$$A_m^{\mp} = k(x, m) \mp \frac{d}{dx} \tag{65}$$

and where $k(x, m) = \alpha m \tanh \alpha x$; also $\epsilon_m = \alpha^2 m^2$, and n is the eigenvalue index,

$$\Psi_n = \psi_{m-n}^m, \quad E_n = -\epsilon_{m-n} = -\alpha^2 (m-n)^2, \quad n = 0, 1, 2 \ldots < m. \tag{66}$$

Beginning with the zeroth-order eigenfunctions, the eigenfunctions can be found by successive applications of the raising operator, which only increases the value of the upper index. That is,

$$\psi_\ell^\ell(x) = \sqrt{\frac{\alpha\Gamma(\ell + \frac{1}{2})}{\sqrt{\pi}\Gamma(\ell)}} \cosh^{-\ell} \alpha x. \tag{67}$$

We repeatedly apply the creation operator $A_{s+1}^- \psi_\ell^s = \psi_\ell^{s+1}$. Note that from (63) and (64), $A_m^- A_m^+$ and $A_m^+ A_m^-$ give different Hamiltonian operators.

4.1 Two-parameter factorization of the Pöschl-Teller Hamiltonian

Following our previous work [2, 3], we define two non-mutually adjoint first-order operators,

$$B_m = \eta_m^{-1}\frac{d}{dx} + \beta_m, \qquad B_m^* = -\eta_m \frac{d}{dx} + \beta_m, \tag{68}$$

where β_m and η_m are functions of x, and we require that $B_{m+1}B_{m+1}^* = H_{m+1} + \epsilon_{m+1}$. Then β_{m+1} and η_{m+1} are the solutions of

$$-\frac{\eta'}{\eta} + \frac{\beta}{\eta} - \beta\eta = 0, \qquad \frac{\beta'}{\eta} + \beta^2 = -\frac{\alpha^2 m(m+1)}{\cosh^2 \alpha x} + \epsilon. \tag{69}$$

By multiplying the first equation by β/η and adding, we have that

$$\left(\frac{\beta_{m+1}}{\eta_{m+1}}\right) + \left(\frac{\beta_{m+1}}{\eta_{m+1}}\right)^2 = -\frac{\alpha^2 m(m+1)}{\cosh^2 \alpha x} + \epsilon_{m+1}. \tag{70}$$

This Riccati equation was found in [24]; it has the solution $\beta/\eta = D \tanh \alpha x$, with $\epsilon = D^2$, and two possible values for D, $D = \alpha(m+1), \; -\alpha m$. If we simply set $\eta_m \to 1$, we recover the factorization (63).

The constant ϵ is usually related to the lowest energy eigenvalue, but here the two different values come from the index asymmetry in the factorizations (63) and (64). Following Ref. [24], we solve for $D = \alpha(m+1)$.

The general solution to the pair of coupled equations (69) is

$$\eta_{m+1}(x) = \left[1 + \frac{\gamma_2 \operatorname{sech}^{2(m+1)} \alpha x}{\left(1 + \gamma_1 \int_0^x \operatorname{sech}^{2(m+1)} \alpha y \, dy\right)^2}\right]^{-1/2}, \tag{71}$$

and

$$\beta_{m+1}(x) = \left[\alpha(m+1) \tanh \alpha x + \frac{\gamma_1 \operatorname{sech}^{2(m+1)} \alpha x}{1 + \gamma_1 \int_0^x \operatorname{sech}^{2(m+1)} \alpha y \; dy}\right] \times \eta_{m+1}(x), \tag{72}$$

where γ_1 has to satisfy $|\gamma_1| < 2\alpha\Gamma(m + 3/2)/(\sqrt{\pi}\Gamma(m+1))$. The corresponding condition on γ_2 involves transcendental functions, but one may use $\gamma_2 > -1 + \gamma_1^2$ to determine the (γ_1, γ_2) parameter space. When $\gamma_1 = \gamma_2 = 0$ we recover the original IH raising/lowering operators.

4.2 Reversing the operator product: new Sturm-Liouville operator

Now we invert the first-order operators' product, keeping in mind Eq. (64),

$$B_m^* B_m = -\frac{d^2}{dx^2} + 2\frac{\eta_m'}{\eta_m}\frac{d}{dx} + \left(V_0 + \epsilon_m - \eta_m \beta_m' - \frac{\beta_m'}{\eta_m}\right). \tag{73}$$

Then we can define a new Sturm-Liouville (SL) eigenvalue problem $\mathcal{L}\Phi_n + \omega(x)E_n\Phi_n = 0$, where

$$\mathcal{L} = \frac{d}{dx}\left[\eta_m^{-2}\frac{d}{dx}\right] + \left(\epsilon_m - \beta_m^2\right)\left(1 + \eta_m^{-2}\right) - \alpha^2 m(m+1)\,\text{sech}^2(\alpha x) \tag{74}$$

$$\Phi_n = \phi_{m-n}^m \equiv B_m^* \psi_{m-n}^{m-1}, \tag{75}$$

with the weight function $\omega(x) = \eta_m^{-2}(x)$.

This new SL operator is isospectral to the original PT problem. The zeroth-order eigenfunction is easily found by solving $B\phi_0 = \left[\frac{d}{dx} + \beta_m\eta_m\right]\Phi_0 = 0$ which gives

$$\Phi_0 = \eta_m(x) \times \frac{\text{sech}^{m+1}(\alpha x)}{1 + \gamma_1\int_0^x \text{sech}^{2(m+1)}(\alpha y)dy}. \tag{76}$$

4.3 Regions in the two-parameter space

We may recover the original QM problem when $\gamma_1 = \gamma_2 = 0$, the origin of the two-parameter space. Moreover, the SUSY partner of the PT problem arises when one sets $\gamma_2 = 0$, moving along the horizontal axis. In this case, $\mathcal{L}$ becomes

$$\mathcal{L} = \frac{d^2}{dx^2} + \alpha^2\lambda(\lambda+1)\,\text{sech}^2(\alpha x) - 2S_1^2(\alpha x) - 4\alpha\lambda\tanh(\alpha x)S_1(\alpha x) \tag{77}$$

where $\lambda = m + 1$, with $S_1(\alpha x) = \frac{\gamma_1\,\text{sech}^{2\lambda}(\alpha x)}{1+\gamma_1\int_0^x \text{sech}^{2\lambda}\alpha y\,dy}$, and $\omega(x) = 1$. These in turn define a SUSY PT problem

$$\left[-\frac{d^2}{dx^2} + \tilde{V}(x)\right]\Phi_n = E_n\Phi_n(x) \tag{78}$$

where the partner SUSY potentials are given by

$$\tilde{V} = -\alpha^2\lambda(\lambda+1)\,\text{sech}^2(\alpha x) + 2S_1^2(\alpha x) + 4\alpha\lambda\tanh(\alpha x)S_1(\alpha x). \tag{79}$$

The zeroth-order eigenfunction is defined by $B^-\phi_0 = 0$, that is,

$$\phi_0 = \frac{\text{sech}^{\lambda}(\alpha x)}{1 + \gamma_1\int_0^x \text{sech}^{2\lambda}(\alpha y)dy}. \tag{80}$$

5. Quasi-exactly solvable potentials

In exactly solvable problems, the whole spectrum is found analytically, but the vast majority of problems have to be solved numerically. A new possibility arises with the class of QES potentials, where a subset of the spectrum may be found analytically [25–27]. QES potentials have been studied using the Lie algebraic method [25]: Manning [28], Razavy [29], and Ushveridze [30] potentials belong to this class (see also [31]). These are double-well potentials, which received much attention due to their applications in theoretical and experimental problems. Furthermore, hyperbolic-type potentials are found in many physical applications, like the Rosen-Morse potential [32], Dirac-type hyperbolic potentials [33], bidimensional quantum dot [34], Scarf-type entangled states [35], etc. QES potentials' classification has been given by Turbiner [25] and Ushveridze [30].

Here we show that the Lie algebraic procedure may impose strict restrictions on the solutions: we shall construct here analytical solutions for the Razavy-type potential $V(x) = V_0(\sin h^4(x) - k \sin h^2(x))$ based on the polynomial solutions of the related confluent Heun equation (CHE) and show that in that case the energy eigenvalues diverge when $k \to -1$, a feature solely of the procedure. We shall also show that other QES potentials may be found that do not belong to any of the potentials found using the Lie algebraic method.

5.1 A Razavy-type QES potential

Let us consider Schrödinger's problem for the Razavy-type potential $V(x) = V_0(\sin h^4(x) - k \sin h^2(x))$,

$$\frac{-\hbar^2}{2\mu}\frac{d^2\psi(x)}{dx^2} + V_0(\sinh^4(\lambda x) - k \sin h^2(\lambda x))\psi(x) = E\psi(x). \tag{81}$$

For simplicity, we set $\mu = \hbar = \lambda = 1$ [35, 36].

Here the potential function is the hyperbolic Razavy potential $V(x) = \frac{1}{2}(\zeta \cosh(2x) - M)^2$, with $V_0 = 2\zeta^2$, where M energy levels are found if M is a positive integer [29]. It may also be viewed as the Ushveridze potential $V(x) = 2\xi^2 \sinh^4(x) + 2\xi[\xi - 2(\gamma + \delta) - 2\ell]\sinh^2(x) + 2(\delta - \frac{1}{4})(\delta - \frac{3}{4})\operatorname{csch}^2(x) - 2(\gamma - \frac{1}{4})(\gamma - \frac{3}{4})\operatorname{sech}^2(x)$, when $\gamma = \frac{1}{4}$ and $\delta = \frac{3}{4}$, or vice versa [30], which is QES if $\ell = 0, 1, 2, \cdots$ (with $\delta \geq \frac{1}{4}$). El-Jaick et al. showed that it is also QES if ℓ =half-integer and $\gamma, \delta = \frac{1}{4}, \frac{3}{4}$ [37].

In the case of the Razavy potential, the solutions obtained by Finkel et al. are

$$\psi_{\sigma\eta}(x, E_R) \propto (\sinh x)^{\frac{1}{2}(1-\sigma-\eta)}(\cosh x)^{\frac{1}{2}(1-\sigma+\eta)} e^{-\frac{\zeta}{2}\cosh(2x)} \sum_{j=0}^{n} \frac{\hat{P}_j^{\sigma\eta}(E_R)}{\left(2j + \frac{\eta-\sigma+1}{2}\right)!} \cosh^{2j}(x) \tag{82}$$

with the parameters $(\sigma, \eta) = (\pm 1, 0)$ or $(0, \pm 1)$, the energy eigenvalues being the roots of the polynomials $P_{j+1}^{\sigma\eta}(E_R)$, satisfying the three-term recursive relations

$$\hat{P}_{j+1}^{\sigma\eta} = (E_R - b_j)\hat{P}_j^{\sigma\eta}(E_R) - a_j \hat{P}_{j-1}^{\sigma\eta}(E_R), \qquad j \geq 0 \tag{83}$$

with $E_R = 2E$, and

$$\begin{aligned} a_j &= 16\zeta j(2j - \sigma + \eta)(j - n - 1) \\ b_j &= -4j(j + 1 - \sigma + 2\zeta) + (2n + 1)(2(n - \sigma) + 3) + \zeta(\zeta - 2\eta + 4n). \end{aligned} \tag{84}$$

5.2 Symmetric solutions for $V(x) = V_0 \sinh^4(x)$

To find the even solutions to Eq. (81) with $k = 0$, let us set $\beta(x) = \cosh^2(x)$, to get

$$\beta(\beta - 1)\frac{d^2\psi}{d\beta^2} + \left(\beta - \frac{1}{2}\right)\frac{d\psi}{d\beta} + \frac{1}{4}\left[2E - 2V_0\beta^2 + 4V_0\beta - 2V_0\right] = 0 \tag{85}$$

and to ensure that $\psi(x)$ vanishes as $x \to \pm\infty$, let $\psi(x) = e^{-\frac{\alpha\beta}{2}}f(\beta)$. Previous works may not include square-integrable solutions to the Razavy potential [38–40]. By requiring $\alpha^2 = 2V_0$, we obtain [41]

$$\beta(\beta - 1)\frac{d^2f}{d\beta^2} + \left[-\alpha\beta(\beta - 1) + \left(\beta - \frac{1}{2}\right)\right]\frac{df}{d\beta} + \left[\frac{\alpha^2\beta}{4} - \frac{\alpha\beta}{2} + \frac{\alpha}{4} + \frac{E}{2} - \frac{\alpha^2}{4}\right]f = 0. \tag{86}$$

We shall look for rank N polynomial solutions: $f(\beta) = f_0$ for $N = 0$, or $f(\beta) = f_0 \prod_{i=1}^{N}(\beta - \beta_i)$ for $N > 0$, the β_i being the roots of the resulting polynomial in Eq. (86). Sometimes the $N = 0$ solution is not even considered [35].

The highest power of β in Eq. (86) fix α to $\alpha = 4N + 2$. The energy eigenvalues and the roots satisfy

$$E = \frac{1}{2}\left[\alpha^2 + \alpha\left(4\sum_{i=1}^{N}\beta_i - 1 - 4N\right) - 4N^2\right] \tag{87}$$

$$\sum_{i \neq j}^{N}\frac{2}{\beta_i - \beta_j} + \frac{-\alpha\beta_i^2 + (\alpha + 1)\beta_i - \frac{1}{2}}{\beta_i^2 - \beta_i} = 0, \qquad i = 1, 2, ..., n. \tag{88}$$

V_0 is found to depend on the order of the polynomial, $V_0 = 2(2N + 1)^2$ for even solutions, and solutions with different N cannot be scaled one into the other due to the $\sinh^4(x)$ dependence of the potential function. The highest solution order is $n = 2N$, and we use subindexes $\{N, n\}$ to label eigenvalues/eigenfunctions.

For $N = 0, f(\beta) = 1$, we get $V_0 = 2$, $E_{0,0} = 1$, and the (unnormalized) ground-state eigenfunction $\psi_{0,0}(x) = e^{-\cosh^2(x)}$. For $N = 2, f(\beta) = f_0(\beta - \beta_1)(\beta - \beta_2)$, equating to zero the coefficients of the polynomial $P(\beta)$, we get the coupled equations

$$\begin{aligned} &\frac{\alpha^2}{4} - \frac{5\alpha}{2} = 0 \\ &3 + (\beta_1 + \beta_2)\left(-\frac{\alpha^2}{4} + \frac{3\alpha}{2}\right) + \left(-\frac{\alpha^2}{4} + \frac{9\alpha}{2} + \frac{E}{2}\right) = 0 \\ &-3 - (\beta_1 + \beta_2)\left(-\frac{\alpha^2}{4} + \frac{5\alpha}{4} + \frac{E}{2} + 1\right) + \beta_1\beta_2\left(\frac{\alpha^2}{4} - \frac{\alpha}{2}\right) = 0 \\ &\frac{1}{2}(\beta_1 + \beta_2) + \beta_1\beta_2\left(-\frac{\alpha^2}{4} + \frac{\alpha}{4} + \frac{E}{2}\right) = 0. \end{aligned} \tag{89}$$

Solving these, we find that $V_0 = 50$, and the three possible eigenvalues, $E_{2,0} = 2.6301$, $E_{2,2} = 19.0121$, and $E_{2,4} = 43.2490$.

5.3 Antisymmetric solutions

In order to find antisymmetric solutions to Eq. (86), we set $f(\beta) = \sinh(x)g(\beta)$, to obtain

$$\begin{aligned}\beta[\beta-1]\frac{d^2g}{dx^2} + \left[-\alpha\beta^2 + (\alpha+2)\beta - \frac{1}{2}\right]\frac{dg}{dx} \\ + \left[\left(-\alpha + \frac{\alpha^2}{4}\right)\beta + \left(-\frac{\alpha^2}{4} + \frac{\alpha}{4} + \frac{E}{2} + \frac{1}{4}\right)\right]g = 0.\end{aligned} \tag{90}$$

This CHE can be solved in power series: $g(\beta) = g_0$ if $N = 0$, or $g(\beta) = g_0 \prod_{i=1}^{N} (\beta - \beta_i)$ for $N > 0$. Then, $\alpha = 4(N+1)$, and

$$E = \frac{1}{2}\left[\alpha^2 + \alpha\left(4\sum_{i=1}^{N}\beta_i - 1 - 4N\right) - 4N^2 - 4N - 1\right]. \tag{91}$$

Here, $V_0 = 8(N+1)^2$, and all even and odd solutions have different V_0. The maximum solutions order is $n = 2N+1$. For example, for $N = 3$ we get $\alpha = 16$, $V_0 = 128$, and

$$\begin{aligned}&(\beta_1+\beta_2+\beta_3)\left(3\alpha - \frac{\alpha^2}{4}\right) + \left(-\frac{\alpha^2}{4} + \frac{13\alpha}{4} + \frac{E}{2} - \frac{49}{4}\right) = 0\\ &(\beta_1+\beta_2+\beta_3)\left(\frac{\alpha^2}{4} - \frac{9\alpha}{4} - \frac{E}{2} - \frac{25}{4}\right) + (\beta_1\beta_2+\beta_2\beta_3+\beta_3\beta_1)\left(\frac{\alpha^2}{4} - 2\alpha\right) - \frac{15}{2} = 0\\ &3(\beta_1+\beta_2+\beta_3) + (\beta_1\beta_2+\beta_2\beta_3+\beta_3\beta_1)\left(-\frac{\alpha^2}{4} + \frac{5\alpha}{4} + \frac{9}{4} + \frac{E}{2}\right) + \beta_1\beta_2\beta_3\left(-\frac{\alpha^2}{4} + \alpha\right) = 0\\ &-\frac{1}{2}(\beta_1\beta_2+\beta_2\beta_3+\beta_3\beta_1) - \beta_1\beta_2\beta_3\left(\frac{\alpha^2}{4} - \frac{\alpha}{4} - \frac{E}{2} - \frac{1}{4}\right) = 0.\end{aligned} \tag{92}$$

We find four eigenvalues, $E_{3,1} = 12.8152$, $E_{3,3} = 40.4568$, $E_{3,5} = 75.7246$, and $E_{3,7} = 117.003$.

6. The potential function $V(x) = V_0(\sinh^4(x) - k\sinh^2(x))$

Now we apply our analysis to the problem with $V(x) = V_0(\sinh^4(x) - k \sinh^2(x))$, which is a symmetric double well if $k > 0$. To find even solutions, we set again $\beta(x) = \cosh^2(x)$ and $\psi(\beta) = e^{-\frac{\alpha}{2}\beta}f(\beta)$, with $\alpha^2 = 2V_0$,

$$\beta(\beta-1)\frac{d^2f}{d\beta^2} + \left[-\alpha\beta(\beta-1) + \left(\beta - \frac{1}{2}\right)\right]\frac{df}{d\beta}$$

$$+\left[\frac{\alpha^2\beta}{4}(1+k) - \frac{\alpha\beta}{2} + \frac{\alpha}{4} + \frac{E}{2} - \frac{\alpha^2}{4}(1+k)\right]f = 0. \tag{93}$$

We now find that $V_0 = \frac{2(2N+1)^2}{1+k}$, k varying freely. For example, if $N = 0$, $E_{0,0} = 1/(1+k)$, and no negative energy eigenvalues may exist. For $N = 1$ the two energy eigenvalues found are

$$E = \frac{9 - (1+k) \pm \sqrt{(1+k)^2 + 36}}{1+k} \tag{94}$$

meaning that for $k > 3/2$ we will have negative eigenvalues. Note that for $N > 0$, it is always possible to find a zero-energy ground state, a feature that may have cosmological implications [18].

For the case with $N = 2$, choosing $k = 4$, the energy eigenvalues are $E_{2,0} = -3.74456$, $E_{2,2} = 1.00000$, and $E_{2,4} = 7.74456$. The corresponding eigenfunctions are plotted in **Figure 1**.

Now, to find the antisymmetric eigenfunctions, we set $f(\beta) = \sinh(x)\ g(\beta)$, to get the CHE

$$\beta(\beta-1)\frac{d^2g}{d\beta^2} + \left[-\alpha\beta^2 + (\alpha+2)\beta - \frac{1}{2}\right]\frac{dg}{d\beta}$$
$$+\left[\beta\left(\frac{\alpha^2}{4}(1+c) - \alpha\right) + \left(\frac{\alpha}{4} + \frac{E}{2} - \frac{\alpha^2}{4}(1+c) + \frac{1}{4}\right)\right]g = 0\ . \tag{95}$$

For $N = 0$ we get that $\alpha = 4/(1+k)$ and $E_1 = 6/(1+k) - 1/2$, such that if $k > 11$, we may find negative energy eigenvalues. For $N = 2$, $\alpha = 12/(1+k)$, if we set $k = 5$, the energy eigenvalues found are $E_{2,1} = -7.11693$, $E_{2,3} = 1.08119$, and $E_{2,5} = 9.53574$. The eigenfunctions are plotted in **Figure 1**.

Note that in this case $(E_1 - E_0)/E_0 = 0.0052$, and it is not possible to distinguish these eigenvalue's lines from each other in **Figure 1** for antisymmetric

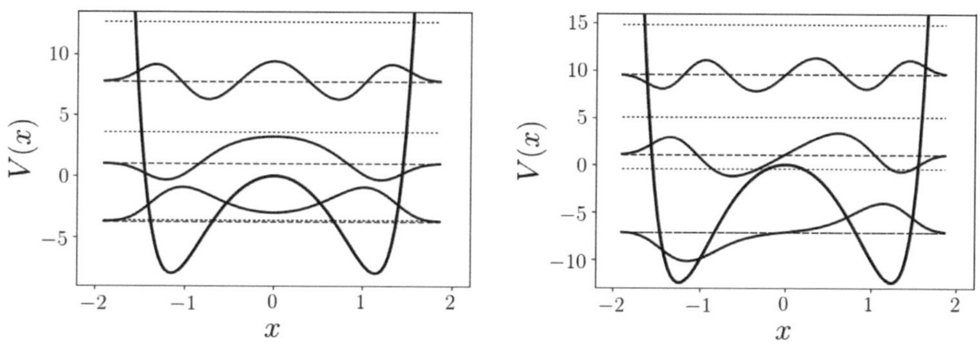

Figure 1.
Left: the three even eigenfunctions (narrow solid lines) found analytically for k = 4 and N = 2, together with the corresponding eigenvalues (dashed lines). Right: the three odd eigenfunctions (narrow solid lines) found analytically for k = 5 and N = 2, together with the corresponding eigenvalues (dashed lines). The unsolved eigenvalues are shown in dotted lines.

eigenvalues, implying quasi-degenerate eigenstates. A similar effect is seen in the symmetric case.

6.1 The case with $k = -1$

As was seen in Section VI, the ground-state energy diverges as $1/(1+k)$ and as $k \to -1$, and this also happens to all higher-order even eigenvalues (see Eq. (94)). This is a strange behavior, since it is clear that the potential function has a rather simple functional form for any value of k: a single or double well with infinite barriers. We can see that this is only a characteristic due to the analytical solution procedure, coming from the fact that the potential strength V_0 is also divergent when $k \to -1$.

6.2 Unclassified QES potentials

Finally, we would like to emphasize that there should be other potential functions which may not be classified form the Lie algebraic method [25].

Indeed, let us consider Schrödinger's problem with the potential function

$$V(x) = \frac{\alpha^2}{2}\cosh^2(x) - \frac{3\alpha}{2}\cosh(x) + \frac{\alpha}{\cosh(x)}. \tag{96}$$

For this problem, the ground-state eigenfunction and eigenvalue are given by

$$\psi = \psi_0 e^{-\alpha \cosh(x)} \cosh(x), \qquad E = \frac{\alpha^2 - 1}{2} \tag{97}$$

while this particular problem does not belong to the class of potentials found using the Lie algebraic method. Similar potentials may be found which do not belong to that class, leaving space for further developments.

Acknowledgements

This work was partially supported by CONACYT 179881 grants and PROMEP grants UGTO-CA-3. This work is part of the collaboration within the Instituto Avanzado de Cosmología. E. Condori-Pozo is supported by a CONACYT graduate fellowship.

Author details

José Socorro García Díaz*, Marco A. Reyes, Carlos Villaseñor Mora and Edgar Condori Pozo
Departamento de Física, División de Ciencias e Ingenierías, Universidad de Guanajuato-Campus León, León, Guanajuato, Mexico

*Address all correspondence to: socorro@fisica.ugto.mx

References

[1] Cooper F, Khare USA. Supersymmetry in Quantum Mechanics. Singapore: World Scientific; 2001

[2] Reyes MA, Rosu HC, Gutiérrez MR. Self-adjoint oscillator operator from a modified factorization. Physics Letters A. 2011;**375**:2145-2148

[3] Arcos-Olalla R, Reyes MA, Rosu HC. An alternative factorization of the quantum harmonic oscillator and two-parameter family of self-adjoint operators. Physics Letters A. 2012;**376**: 2860-2865

[4] Darboux G. Analyse Mathématique. Sur une proposition relative aux équations linéaires. Comptes Rendus de l'Académie des Sciences. 1882;**94**:1456

[5] Ince EL. Ordinary Differential Equations. New York: Dover; 1926

[6] Matweev VB, Salle MA. Darboux Transformation and Solitons. Berlin: Springer; 1991

[7] Cooper F, Khare A, Sukhatme U. Supersymmetry and quantum mechanics. Physics Reports. 1995; **251**:267

[8] Mielnik B. Factorization method and new potentials with the oscillator spectrum. The Journal of Mathematical Physics. 1984;**25**:3387

[9] Nieto MM. Relationship between supersymmetry and the inverse method in quantum mechanics. Physics Letters B. 1984;**145**:208

[10] Fernández DJ. New hydrogen-like potentials. Letters in Mathematical Physics. 1984;**8**:337

[11] Graham R. Supersymmetric Bianchi type IX cosmology. Physical Review Letters. 1991;**67**:1381

[12] Obregón O, Socorro J, Benítez J. Supersymmetric Quantum Cosmology proposals and the Bianchi type II Model. Physical Review D. 1993;**47**:4471

[13] Lidsey JE. Scale factor duality and hidden supersymmetry in scalar-tensor cosmology. Physical Review D. 1995;**52**: R5407

[14] Socorro J, Medina ER. Supersymmetric Quantum Mechanics for Bianchi Class A Models. Physical Review D. 2000;**61**:087702

[15] Obregón O, Rosales JJ, Socorro J, Tkach VI. Supersymmetry breaking and a normalizable wave function for the FRW (k=0) cosmological model. Classical and Quantum Gravity. 1999; **16**:2861

[16] Socorro J. Supersymmetric quantum solution for FRW cosmological model with matter. Revista Mexicana de Física. 2002;**48**(2):112

[17] Socorro J, Obregón O. Quantum cosmology for inflationary scenary. Revista Mexicana de Física. 2002;**48**(3):205

[18] Socorro J, Nuñez OE. Scalar potentials with multi-scalar fields from quantum cosmology and supersymetric quantum mechanics. The European Physical Journal—Plus. 2017;**132**:168

[19] Filho ED. Supersymmetric solution for 2–dimensional Schrodinger equation. Modern Physics Letters A. 1993;**8**(1):63

[20] Witten E. Dynamical breaking of supersymmetry. Nuclear Physics B. 1981;**188**:513

[21] Faddeev LD, Salvnov AA. Gauge Fields: An Introduction to Quantum Theory. Reading, MA: Addison-Wesley; 1991. sec 2.5

[22] Rosen N, Morse PM. On the vibrations of polyatomic molecules. Physical Review. 1932;**42**:210-217

[23] Infeld L, Hull TE. The factorization method. Reviews of Modern Physics. 1951;**23**:21-68

[24] Díaz JI, Negro J, Nieto LM, Rosas-Ortiz O. The supersymmetric modified Pschl-Teller and delta well potentials. Journal of Physics A: Mathematical. 1999;**32**:8447-8460

[25] Turbiner AV. Quasi-exactly-solvable problems and sl(2) algebra. Communication in Mathematical Physics. 1988;**118**:467

[26] Shifman MA. New findings in quantum mechanics (Partial algebraization of the spectral problem). International Journal of Modern Physics A. 1989;**04**(12):2897

[27] Ushveridze AG. Soviet Journal of Nuclear Physics. 1989;**20**:504

[28] Xie Q-T. New quasi-exactly solvable double-well potentials. Journal of Physics A: Mathematical and Theoretical. 2012;**45**:175302

[29] Razavy M. An exactly soluble Schrdinger equation with a bistable potential. American Journal of Physics. 1980;**48**:285-288

[30] Ushveridze AG. Quasi-Exactly Solvable Models in Quantum Mechanics. Briston: Institute of Physics; 1993

[31] Chen BH, Wu Y, Xie Q-T. Heun functions and quasi-exactly solvable double-well potentials. Journal of Physics A: Mathematical and Theoretical. 2013;**46**:035301

[32] Oyewumi KJ, Akoshile CO. Bound-state solutions of the Dirac-Rosen-Morse potential with spin and pseudospin symmetry. European Physical Journal A. 2010;**45**(3):311

[33] Wei GF, Liu XY. The relativistic bound states of the hyperbolical potential with the centrifugal term. Physica Scripta. 2008;**78**:065009

[34] Xie WF. Singlet-Triplet Transitions of a Pschl-Teller Quantum Dot. Communications in Theoretical Physics. 2006;**46**:1101

[35] Downing CA. On a solution of the Schrdinger equation with a hyperbolic double-well potential. Journal of Mathematical Physics. 2013;**54**:072101

[36] Wen FK, Yang ZY, Liu C, Yang WL, Zhang YZ. Exact Polynomial Solutions of Schrdinger Equation with Various Hyperbolic Potentials. Communications in Theoretical Physics. 2014;**61**:153-159

[37] El-Jaick LJ, Figuereido BDB. Confluent Heun equations: Convergence of solutions in series of coulomb wavefunctions. Journal of Physics A. 2013;**48**:085203

[38] Khare A, Mandal BP. Anti-isospectral transformations, orthogonal polynomials, and quasi-exactly solvable problems. Journal of Mathematical Physics. 1998;**39**:3476

[39] Konwent H, Machnikowsky P, Magnuszelwski P, Radosz A. Some properties of double-Morse potentials. Journal of Physics A: Mathematical and General. 1998;**31**:7541

[40] Finkel F, Gonzalez-Lopez A, Rodriguez MA. On the families of orthogonal polynomials associated to the Razavy potential. Journal of Physics A: Mathematical and General. 1999;**32**: 6821

[41] Zhang Y. Exact polynomial solutions of second order differential equations and their applications. Journal of Physics A: Mathematical and Theoretical. 2012;**45**:065206